DEFORMATION AND STRENGTH
OF MATERIALS

DEFORMATION AND STRENGTH OF MATERIALS

P. FELTHAM

D.Sc., F.Inst.P.

Brunel University, London W.3

Springer Science+Business Media, LLC

1966

Suggested U.D.C. number : 539·3/·4

Originally published by Butterworths in 1966.
Softcover reprint of the hardcove 1st edition 1966

ISBN 978-1-4899-5849-5   ISBN 978-1-4899-5847-1 (eBook)
DOI 10.1007/978-1-4899-5847-1

PREFACE

ONE of the effects of the unprecedented advance in the synthesis and use of new materials in the last few decades, but particularly since the end of the Second World War, has been the widespread introduction of 'materials science' into curricula of university courses. The momentum persists, and from the point of view of research and applications a vast array of properties of alloys, ceramics, semiconductors, polymers, fluids, cellular and other materials remains in the focus of interest.

How to distil from this embarrassing wealth of matter the essential features necessary for an adequate interpretation of their properties, how in fact to introduce the theoretical foundations of materials science into university curricula in an effective yet concise manner has therefore been a topical problem for some time.

In the present textbook, which is based largely on lectures to science and technology undergraduates, I have attempted the task of dealing systematically with the aspects of materials science which, particularly in the light of my experience in industrial research, appeared to me to be essential to an understanding of the fundamental fabric of principles. The boundaries of the subject are not well defined; a study of materials must have its roots in the firm soil of the classical sciences. An elementary knowledge of mathematics has therefore been assumed. Topics well covered in standard texts on physical chemistry, as well as problems involving electron transport, which are ably dealt with in many books on 'solid state' have not been included. The emphasis has been on the mechanical properties of materials, including elasticity, visco-elastic behaviour, damping capacity, the strength of real crystals, dislocation theory, fracture, fatigue, and the behaviour of non-Newtonian fluids.

Wherever possible I have tried to show the connection between different forms of behaviour, and to deduce quantitative relations in the simplest possible way. Although the theoretical foundations of macroscopic elasticity and plasticity are well established, the micromechanisms of plastic flow and fracture, for example, have not as yet been fully elucidated in many materials, and they may in fact be subjects of keen controversy. This state of affairs is responsible for the tentative approach which the reader will occasionally find in the book, for example in the treatment of work-hardening.

The book is aimed at a large circle of readers interested in the theoretical foundations of the subject, and it should prove particularly useful to students in the science, technology and medical faculties of universities. The emphasis has therefore been on the clear and concise explanation of the essential, basic, concepts of the mechanical properties of materials, and not on the detailed survey of the properties of the great variety solids and liquids encountered in practice. Its function is considered to be an exposition which should enable the serious reader to learn the language of the subject. Thus armed he can then search with profit in papers and specialised handbooks for guidance on more specific problems.

London   P.F.

CONTENTS

1

ELASTICITY

1.1 Introductory

Materials which can be subjected to significant deformations, but which regain their original shape when the constraints are removed, are solids. This complete recovery of form, observed at least up to certain levels of the stress, characterising the elastic range, distinguishes them from liquids, which will suffer irreversible changes of shape in any general deformation, however small.

Experience shows that within the elastic range the relation between the deformations and the applied stresses is linear; the relationship is known as Hooke's Law. Beyond the elastic range the material acquires a permanent set; a plastic deformation is said to remain when the forces are removed.

Figure 1.1 shows a stress-strain curve typical of many materials. It can be obtained by applying a steadily increasing tensile stress σ to a rod or wire, and plotting the tensile strain ϵ in the course of deformation. The elastic range terminates at σ_e, while gross plastic yielding occurs above the yield stress σ_y. If at a certain point P the stress is removed, and the test is subsequently continued, the specimen would behave elastically as indicated by the broken line AP, and would then extend from P onward as if no interruption had taken place. The total strain can be seen to consist of the plastic component, OA, and the elastic contribution AB. Young's modulus is given by $\tan \alpha$, which represents the slope of the curve in the elastic range.

The stress at the point P is termed the flow stress at the

1

corresponding strain, to distinguish it from the yield stress which is measured at the initial transition to the plastic state. As this transition may not be sufficiently abrupt to enable one to ascribe a fairly exact value to σ_y a proof stress is generally specified instead in engineering practice, particularly with metals. This represents the value of the flow stress at some small total tensile strain, such as $0 \cdot 1$ or $0 \cdot 2$ per cent, and provides a practical measure of the load-bearing capacity of the material. In general it is only

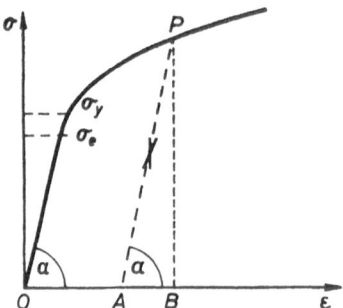

Figure 1.1. Stress-strain curve of a ductile material

slightly higher than σ_y. The smallness of these reference strains gives an indication of the limited extent of the elastic range in crystalline materials, which contrasts sharply, for example, with rubber-like solids.

Structural changes induced by the deformation result in work-hardening, apparent in the need to apply increasingly higher stresses to strain the material as the tensile test proceeds. These changes eventually lead to fracture. If the total strain ϵ_f attained at the instant of fracture is large compared with the elastic component the material is said to be ductile, otherwise it is regarded as brittle. The strain at fracture is in fact a convenient practical index of ductility.

2

At temperatures relatively high with respect to the melting point (T_m), say above about $0 \cdot 3 \, T_m$ (°K), the work-hardened structure generally becomes unstable. The material softens, anneals, in the course of time, and hence also during the test, and the shape of the stress-strain curve becomes appreciably strain-rate dependent. In an unloading-loading cycle, as described above, the initial and final points on the curve may then no longer coincide, and altogether the practical usefulness of work-hardening curves then becomes limited.

Elastic deformation does not only contribute to the overall change of shape of materials subjected to stresses, but it is also a prerequisite for the occurrence of permanent changes of shape in solids. The question of the quantitative representation of states of stress and strain in elastic materials is therefore of basic importance here.

1.2 Uniaxial Stress

A body which is deformed by the application of a force is automatically subjected to an equal and opposite reaction and, in view of the continuity of the material, a state of stress will exist at all points. In analysing this stress it is sufficient to consider the applied forces explicitly, for whatever generalisations are deduced for them applies equally to the reactions.

Now, a force acting on a small area of a solid can be resolved into two orthogonal components, one acting in the surface, the other in a direction at right-angles to it. The first component, together with its reaction, will give rise to a shear stress, the latter to a tensile stress. Although the force is a vector, the system of stresses induced by it is not. This may be seen quite readily by considering that an element of the surface of a body, sufficiently small to be regarded plane, would be specified in a rectangular co-ordinate system by the direction of its normal which, like the force acting on it, is a vector quantity. The stress,

3

defined as the force per unit area, is therefore represented by the ratio of two vectors, which is not a vector itself, but is known as a tensor. Before considering this question further we shall examine some simple states of stress, beginning with a cylindrical rod of circular cross-section subjected to uniaxial tension by a force F, as shown in *Figure 1.2*.

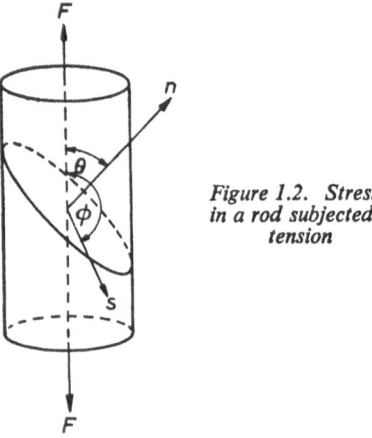

Figure 1.2. Stresses in a rod subjected to tension

The force acting at right-angles to a plane having normal n is $F \cos \theta$ and the area of the ellipse on which it is acting is $A/\cos \theta$, where A is the area of cross-section of the rod. The tensile stress on planes making an angle θ with the axis is therefore $(F/A) \cos^2 \theta$, with a maximum $\sigma = F/A$ on surfaces perpendicular to the rod axis. The tensile stresses may therefore be written as

$$\sigma(\theta) = \tfrac{1}{2}\sigma(1+\cos 2\theta) \qquad (1.1)$$

Similarly, by considering the component $F \sin \theta$ acting on the plane of the ellipse in the direction of its major axis, the shear stress $\tau(\theta)$ is found to be $(F/A) \sin \theta \cos \theta$, or

$$\tau(\theta) = \tfrac{1}{2}\sigma \sin 2\theta \qquad (1.2)$$

4

A geometrical representation of the stresses as function of the angle θ, known as Mohr's circle, is shown in *Figure 1.3*.

It may readily be verified that the co-ordinates of a point on a circle of radius $\frac{1}{2}\sigma$, with centre on the τ-axis and touching the σ-axis are in fact given by equations 1.1 and 1.2 if the radius vector drawn from the centre of the circle to the point P makes an angle 2θ with the shear stress axis. In view of the simplicity of equations 1.1 and 1.2 Mohr's circle is not here of material assistance in the

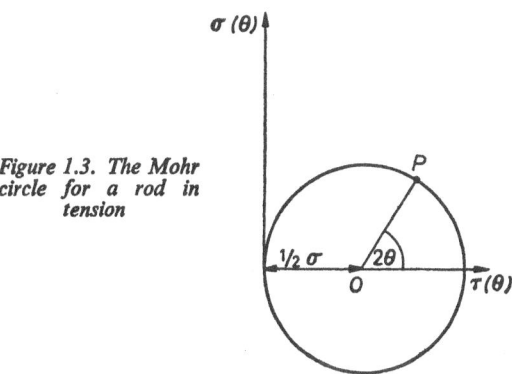

Figure 1.3. The Mohr circle for a rod in tension

determination of stresses, but geometrical methods based on the extension of this principle are frequently used in analyses of more complex systems of stress and strain, and in problems of fracture.

Equation 1.2 implies that the maximum shear stress is equal to $\frac{1}{2}\sigma$, and that it occurs on planes inclined at 45° to the tensile axis. This result has important implications in connection with the mode of plastic deformation of ductile materials, as will be seen later.

1.3 Biaxial Stress

A more complex state of stress results from the simultaneous action of two orthogonal tensile stresses. Close approximations to such states of stress may occur, for example, in stretched membranes or in the shells of thin-walled pressure vessels.

Our considerations will apply irrespective of whether one or both of the stresses are negative, i.e. compressive; results for any specific case are obtained simply by the use of the appropriate signs for the stresses. We shall denote these here by σ_1 and σ_2, and take $\sigma_1 > 0 > \sigma_2$. The diagrams in *Figure 1.4* show that this system can be split into two components, one of which is a uniaxial tension, and the

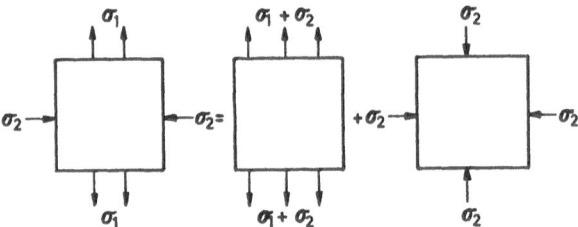

Figure 1.4. Resolution of a biaxial stress

other a two-sided compression. The first will give rise to shear and tensile stresses analogous to those given by equations 1.1 and 1.2, with $\frac{1}{2}(\sigma_1 + \sigma_2)$ substituted for $\frac{1}{2}\sigma$, and, of course, to corresponding changes of shape, in this case to an axial elongation. The second system will not induce any changes in shape of the membrane or foil, except in the thickness (neglected in this case), but with compressive stresses as in *Figure 1.4*, a square lamella would become smaller in size, yet at the same time it would remain square. The system could again be represented by a Mohr circle. The radius would be $\frac{1}{2}(\sigma_1 + \sigma_2)$, and the centre would lie on the τ-axis at a distance $\frac{1}{2}(\sigma_1 + \sigma_2) - \sigma_2$ from the origin.

The system could be split up in other ways, for example by isolating σ_1 rather than σ_2. This arbitrariness disappears

6

if the system is considered to be a special case of a triaxial stress with $\sigma_3 = 0$, as will become apparent below.

1.4 Deviatoric and Isotropic Stresses

In the three-dimensional stress system we shall consider three mutually perpendicular stresses σ_1, σ_2 and σ_3 acting at right-angles to the faces of a small cube of material. We now wish to separate the component which gives rise to changes of shape, termed the deviator, from the isotropic, or hydrostatic, stress responsible only for changes in volume.

Now, writing σ_h for the hydrostatic stress, one may represent the separation formally by

$$(\sigma_1, \sigma_2, \sigma_3) = (\sigma_1 - \sigma_h, \sigma_2 - \sigma_h, \sigma_3 - \sigma_h) + (\sigma_h, \sigma_h, \sigma_h) \quad (1.3)$$

It is not obvious what combination of the three stresses is required to give σ_h, so we shall put

$$\sigma_h = a\sigma_1 + b\sigma_2 + c\sigma_3 \quad (1.4)$$

where a, b and c are yet undetermined constants. Since however none of the three stresses is in any way preferred, a, b and c must be equal, so that it remains to determine the magnitude of a. To this end we consider a specific case in which all three stresses are very nearly equal so that the stress system consists almost entirely of the hydrostatic component. The deviator is then equal to zero, and we must therefore also have

$$(\sigma_1 - \sigma_h) + (\sigma_2 - \sigma_h) + (\sigma_3 - \sigma_h) \approx 0$$

This relation is compatible with equation 1.4 only if

$$\sigma_h = \tfrac{1}{3}(\sigma_1 + \sigma_2 + \sigma_3) \quad (1.5)$$

which is the required result.

It is clear from equations 1.3 and 1.5 that the deviator cannot now be represented by a uniaxial tensile stress, as in *Figure 1.4* ; all three stresses have to be taken into account.

The decomposition of the stress system is of considerable practical importance. It enables one to find the stresses responsible for changes of shape, and to study their effects separately from those introducing isotropic dilatations or compressions. Metals and most inorganic crystals, for example, may for many purposes be regarded as incompressible, and the contribution of the isotropic stress to the total deformation may often be neglected.

1.5 Stresses at a Point in a Body

It may seem that the stress systems so far considered in which, firstly, the stresses were mutually perpendicular and, secondly, shear stresses were not specifically introduced, are not sufficiently general to be of theoretical interest. However, we shall show that the state of stress in, at least, a small volume around any point of an elastically deformed body may be represented by three mutually perpendicular so-called principal stresses, provided the axes of reference are suitably oriented in the body. The required orientation is obvious in most cases occurring in practice, as we shall illustrate by way of an example below; the principal stresses therefore provide a full description of the state of stress over a volume around the chosen point within which the stresses may be considered to be uniform, without abrupt change.

The state of stress close to some point O in the deformed elastic material could be ascertained by a hypothetical experiment in the following way. A set of three axes is imagined in the body, with O as origin. A small tetrahedron is then cut out as indicated in *Figure 1.5*, the base plane ABC having some arbitrary but known orientation given by the angles α, β and γ which the normal makes with the x, y and z axes respectively. The tetrahedron is then imagined to be removed outside the body. As the initial constraints are then no longer acting, the tetrahedron will distort, and tensile and shear stresses must be applied to it

8

faces to restore the previous shape. If these are measured, then the state of stress at the point O is in principle determined. By considering the equilibrium of forces acting on the tetrahedron we shall show that in general three tensile and three shear stresses are required to specify the stress at a point of an elastically isotropic body.

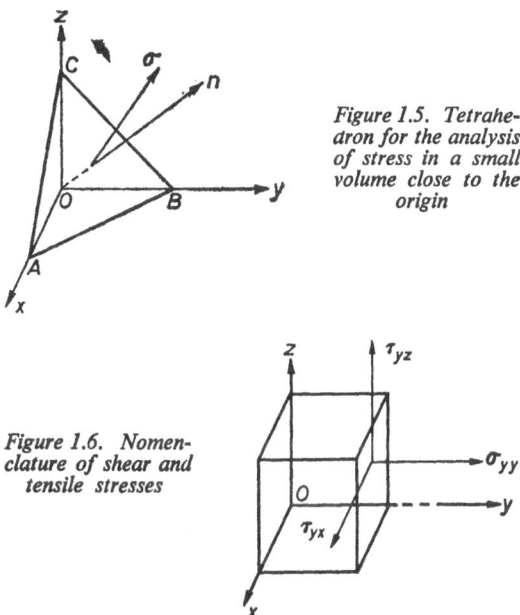

Figure 1.5. Tetrahedron for the analysis of stress in a small volume close to the origin

Figure 1.6. Nomenclature of shear and tensile stresses

Shear stresses will be described by the letter τ with two dissimilar subscripts, tensile stresses by the letter σ with two similar subscripts, as shown for example in *Figure 1.6*. The first subscript specifies the direction of the normal of the plane on, or in, which the stress is acting; for all stresses shown in the figure this is a plane parallel to $y = 0$. The second subscript indicates the direction of the force which gives rise to the stress. This direction is along the normal of the plane in the case of tensile stresses, and both letters of the subscript are then the same. Two different subscripts indicate a shear stress. The nature of the stress is therefore apparent from the subscripts, and the use of two different letters, namely τ and σ, is not essential, although it will be employed here for convenience. A normal stress is tensile or compressive, i.e. negative, depending on whether the two arrows representing it both point away or towards the element on which they act. The convention relating to the sign of shear stresses is as follows.

The couple representing the stress, for example τ_{yz} in *Figure 1.6*, is assumed to act about the origin. Then, if the arrow on the positive side of the origin, in this case at $y > 0$, points along the positive direction of the co-ordinate, i.e. z, the stress is positive; with arrows reversed it is negative.

Now, referring to *Figure 1.5*, we note that the angles which the plane ABC makes with the planes $x = 0$, $y = 0$ and $z = 0$ are α, β and γ respectively, and that consequently the areas of the triangles OBC, OCA and OAB are equal to $\Delta \cos \alpha$, $\Delta \cos \beta$ and $\Delta \cos \gamma$, where Δ is the area of the triangle ABC. Since the forces acting on the tetrahedron are in equilibrium their resolved components in any direction must also be in equilibrium. In particular, if σ_{nx} is the component in the x-direction of the stresses acting on the triangle ABC, then

$$\Delta \sigma_{nx} = \sigma_{xx} \Delta \cos \alpha + \tau_{yx} \Delta \cos \beta + \tau_{zx} \Delta \cos \gamma$$

or

$$\sigma_{nx} = \sigma_{xx} \cos \alpha + \tau_{yx} \cos \beta + \tau_{zx} \cos \gamma$$

and similarly

$$\sigma_{ny} = \tau_{xy} \cos \alpha + \sigma_{yy} \cos \beta + \tau_{zy} \cos \gamma$$

and

$$\sigma_{nz} = \tau_{xz} \cos \alpha + \tau_{yz} \cos \beta + \sigma_{zz} \cos \gamma$$

(1.6)

We now wish to discover whether we could find a plane,

like *ABC* above, on which no shear stresses act, the stresses on *ABC* being purely tensile or compressive. Let us assume that the plane *ABC* in *Figure 1.5* is in fact such a 'principal' plane. The stresses σ_{nx}, σ_{ny} and σ_{nz} are then equal to $\sigma_p \cos \alpha$, $\sigma_p \cos \beta$ and $\sigma_p \cos \gamma$ respectively, where σ_p is the principal stress acting on the triangle *ABC*. On substituting these values into equation 1.6 one obtains

$$\left.\begin{aligned}
(\sigma_{xx} - \sigma_p) \cos \alpha + \tau_{yx} \cos \beta + \tau_{zx} \cos \gamma &= 0 \\
\tau_{xy} \cos \alpha + (\sigma_{yy} - \sigma_p) \cos \beta + \tau_{zy} \cos \gamma &= 0 \\
\tau_{xz} \cos \alpha + \tau_{yz} \cos \beta + (\sigma_{zz} - \sigma_p) \cos \gamma &= 0
\end{aligned}\right\} \quad (1.7)$$

which can be considered as three equations in the unknowns $\cos \alpha$, $\cos \beta$ and $\cos \gamma$, now determining the orientation of the principal plane with respect to the chosen reference axes. As is well known they will yield solutions, in this case real values of σ_p, only if the determinant

$$\begin{vmatrix}
\sigma_{xx} - \sigma_p & \tau_{yx} & \tau_{zx} \\
\tau_{xy} & \sigma_{yy} - \sigma_p & \tau_{zy} \\
\tau_{xz} & \tau_{yz} & \sigma_{zz} - \sigma_p
\end{vmatrix} = 0 \qquad (1.8)$$

On expanding this determinant one obtains a cubic equation in σ_p, yielding three real values σ_1, σ_2 and σ_3, which are the principal stresses at the point *O*. It may readily be shown that they are mutually perpendicular. Hence, from a knowledge of the stresses acting in the planes of an arbitrarily chosen co-ordinate system at *O* the magnitudes of the principal stresses can be determined. If these are substituted in turn into equations 1.7 three sets of direction cosines are obtained. They determine the directions of the principal stresses with respect to the chosen reference system. The state of stress at a point is therefore fully described by six independent data, namely either the magnitude of the three principal stresses and the location of the principal axes at *O* (which necessitates specifying three angles) or by the shear and tensile stresses on the orthogonal faces of the tetrahedron (*Figure 1.5*). Although it may appear from equation 1.8 that nine stress components have to be known to specify the stress, this is not in fact the case, for $\tau_{xy} = \tau_{yx}$, $\tau_{yz} = \tau_{zy}$ and $\tau_{zx} = \tau_{xz}$. This is clear if one considers that the cube in *Figure 1.6* would rotate counterclockwise about an axis through its centre parallel to the x-axis under the action of the shear stress τ_{yz} unless it is balanced by an opposing couple due to a shear stress τ_{zy}, with $\tau_{yz} = \tau_{zy}$.

Equation 1.8, but with σ_p omitted, tabulates the stresses at a point in an isotropic elastic material, and is referred to as the stress tensor; as we have seen, only six of the stresses are independent.

In practice the directions of the principal stresses are often easily identified, as we shall show by way of example below. One then obtains the simple representation of a triaxial stress system considered in section 1.4. By analogy with the uniaxial stress system (equations 1.1 and 1.2) the greatest shear stresses now act on planes inclined 45° to the pairs of stresses (σ_2, σ_3), (σ_3, σ_1) and (σ_1, σ_2) respectively; their absolute values

$$\left.\begin{array}{l} \tau_1 = \frac{1}{2} \left| \sigma_2 - \sigma_3 \right| \\ \tau_2 = \frac{1}{2} \left| \sigma_3 - \sigma_1 \right| \\ \tau_3 = \frac{1}{2} \left| \sigma_1 - \sigma_2 \right| \end{array}\right\} \qquad (1.9)$$

are known as the principal shear stresses.

1.6 Invariants of the Stress Tensor

The cubic equation in σ_p, obtained on expanding the determinant in equation 1.8, may be written formally

$$\sigma_p^3 - I_1 \, \sigma_p^2 - I_2 \, \sigma_p - I_3 = 0 \qquad (1.10)$$

where

$$I_1 = \sigma_{xx} + \sigma_{yy} + \sigma_{zz} \qquad (1.11)$$

and I_2 and I_3 are homogeneous expressions of second and third degrees in the stress components, respectively, as is readily verified. Now, the principal stresses at the point O, which are the roots of equation 1.10, clearly do not depend upon the initial choice of the co-ordinate system at O. Consequently the coefficients in equation 1.10 must also be independent of the choice of axes. It follows that if a differently oriented set of axes had been chosen, for example the principal axes, then one should have

$$\sigma_{xx} + \sigma_{yy} + \sigma_{zz} = \sigma_1 + \sigma_2 + \sigma_3 \qquad (1.12)$$

so that the sum of three mutually perpendicular tensile stresses in any Cartesian co-ordinate system centred on O is equal to I_1 – a constant. Similar conclusions may be drawn for I_2 and I_3. I_1, I_2 and I_3 are known as the first, second and third invariants of the stress tensor of an iso-tropic elastic material. Reference to equation 1.5 shows that

$$I_1 = 3\sigma_h \qquad (1.13)$$

1.7 Elastic Constants and Moduli

Before considering the relation between stress and strain it is necessary to specify the description of strains. As in the case of stresses two types of strain are distinguished, namely shear and tensile.

When dealing with small elastic deformations a tensile strain in any given direction in the body is given by the elongation δl in that direction divided by the initial length l_0. Thus, for a rod with axis along the x-co-ordinate an elongation will give rise to a tensile strain $\epsilon_{xx} = \delta l / l_0$ and, due to the Poisson contraction, to compressive normal strains at right-angles to the axis; thus ϵ_{yy} and ϵ_{zz} will be negative. Referring to the cube of elastic material shown in elevation in *Figure 1.7a*, the shear strain γ_{zy} due to the shear stress τ_{zy} is generally defined in engineering practice by the tangent of the angle AOA', where AA' is the displacement of the upper cube surface. As the strains considered are small the tangent may be replaced by the angle, expressed in radians. The deformation shown in *Figure 1.7a* is characterised by the fact that any plane in the cube parallel

13

to $z = 0$ is displaced rigidly along a straight line, in this case the y-direction, and the separation of any pair of such 'shearing planes' remains constant in the course of deformation; the cube does not extend or contract along the z-axis. Such a shear is referred to as 'simple'. However, as can be seen from *Figure 1.7b* the same change of shape could also be introduced in a symmetric manner by displacing OA and OC in opposite senses such that the angles AOA'' and COC'' are each equal to $\frac{1}{2}\gamma_{zy}$.

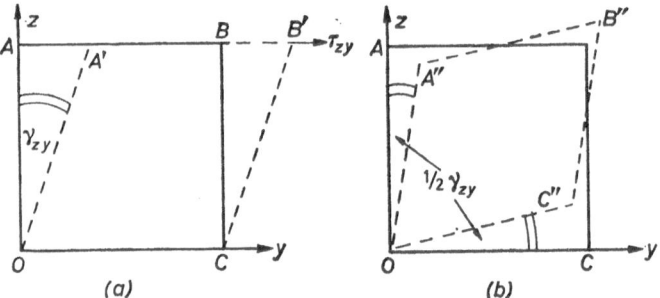

Figure 1.7. (a) *The engineering shear strain;* (b) *rational shear strain*

The angle AOA'' represents the rational shear strain, generally used in the mathematics of elasticity. The deformation which it defines (*Figure 1.7b*) is a 'pure' shear. It describes the change of shape of the elastic cube; the simple shear can be obtained by an additional rotation of the deformed cube through an angle $\frac{1}{2}\gamma_{zy}$ in the clockwise sense about the origin, as is apparent from *Figure 1.7*. A rotation not affecting the shape of the body is therefore implied by the engineering definition of strain, which refers to *Figure 1.7a*.

The principal axes in *Figure 1.7b* lie in the directions $A''C''$, OB'' and the x-axis, and do not change their directions in the course of the deformation. This constancy of the

14

directions of the principal axes is the main feature differentiating a 'pure' from a 'simple' shear. It is readily verified that the principal stresses are

$$(\sigma_1, \sigma_2, \sigma_3) = (-\tau_{zy}, \tau_{zy}, 0) \qquad (1.14)$$

with σ_1 along $A''C''$ and σ_2 along OB''. In this system τ_{zy} is then the principal shear stress τ_3 as defined by equation 1.9. As in the case of stress it may again be shown that six independent strain components are necessary, in general, to specify the state of strain at a point in an elastic, homogeneous, material; the relations $\gamma_{xy} = \gamma_{yx}$, $\gamma_{yz} = \gamma_{zy}$ and $\gamma_{zx} = \gamma_{xz}$ again apply. The strain tensor takes the form

$$\begin{vmatrix} \epsilon_{xx} & \tfrac{1}{2}\gamma_{yx} & \tfrac{1}{2}\gamma_{zx} \\ \tfrac{1}{2}\gamma_{xy} & \epsilon_{yy} & \tfrac{1}{2}\gamma_{zy} \\ \tfrac{1}{2}\gamma_{xz} & \tfrac{1}{2}\gamma_{yz} & \epsilon_{zz} \end{vmatrix}$$

the values of the rational shear strains being used.

In its most general form Hooke's Law may be expressed by representing the six components of the stress at a point by linear combinations of all the strains. One then obtains

$$\sigma_{xx} = c_{11}\epsilon_{xx} + c_{12}\epsilon_{yy} + c_{13}\epsilon_{zz} + c_{14}\gamma_{yz} + c_{15}\gamma_{zx} + c_{16}\gamma_{xy}$$

$$\sigma_{yy} = c_{21}\epsilon_{xx} + c_{22}\epsilon_{yy} + c_{23}\epsilon_{zz} + c_{24}\gamma_{yz} + c_{25}\gamma_{zx} + c_{26}\gamma_{xy}$$

$$\sigma_{zz} = c_{31}\epsilon_{xx} + c_{32}\epsilon_{yy} + c_{33}\epsilon_{zz} + c_{34}\gamma_{yz} + c_{35}\gamma_{zx} + c_{36}\gamma_{xy}$$

$$\tau_{yz} = c_{41}\epsilon_{xx} + c_{42}\epsilon_{yy} + c_{43}\epsilon_{zz} + c_{44}\gamma_{yz} + c_{45}\gamma_{zx} + c_{46}\gamma_{xy}$$

$$\tau_{zx} = c_{51}\epsilon_{xx} + c_{52}\epsilon_{yy} + c_{53}\epsilon_{zz} + c_{54}\gamma_{yz} + c_{55}\gamma_{zx} + c_{56}\gamma_{xy}$$

$$\tau_{xy} = c_{61}\epsilon_{xx} + c_{62}\epsilon_{yy} + c_{63}\epsilon_{zz} + c_{64}\gamma_{yz} + c_{65}\gamma_{zx} + c_{66}\gamma_{xy}$$

The 36 coefficients are known as elastic constants, and are tabulated for various crystalline materials for specific orientations of the stress co-ordinates; these are chosen to coincide, in general, with some of the symmetry axes of the crystal. Only 21 of the coefficients are independent, for it may be shown that $c_{mn} = c_{nm}$, irrespective of whether the material is isotropic or not. This equality is known as Onsager's Reciprocity Relation.

Clearly, it is also possible to express the six components

of strain in terms of the stresses. One then obtains 21 elastic moduli s_{mn}, m, $n = 1, 2, 3$. Each may be expressed in terms of elastic constants, and vice versa.

In the case of elastically isotropic materials, with which we shall be mainly concerned, we shall derive the relation involving the moduli in a representation referred to principal axes. It is of sufficient generality in this form for many applications, and complications due to the explicit introduction of shear stresses and strains are circumvented.

1.8 Hooke's Law referred to Principal Axes

The relation between the principal stresses and the corresponding principal strains ϵ_1, ϵ_2 and ϵ_3 are readily deduced from first principles. We note that the strain due to the application of the stress σ_1 to an isotropic, elastic, cube is σ_1/E, where E is Young's modulus. If now another principal stress is applied, say σ_2, the previous strain will be diminished by $\nu\sigma_2/E$ if σ_2 is positive, or increased by the same amount if σ_2 is a compressive stress. A similar effect would arise from the application of σ_3. Thus, taking Poisson's ratio ν to be positive, one has

$$\left.\begin{aligned}
\epsilon_1 &= \frac{1}{E} \; [\sigma_1 - \nu(\sigma_2 + \sigma_3)] \\[2mm]
\epsilon_2 &= \frac{1}{E} \; [\sigma_2 - \nu(\sigma_3 + \sigma_1)] \\[2mm]
\epsilon_3 &= \frac{1}{E} \; [\sigma_3 - \nu(\sigma_1 + \sigma_2)]
\end{aligned}\right\} \qquad (1.15)$$

If this form of Hooke's Law is compared with the general representation, involving equations such as

$$\epsilon_{xx} = s_{11}\sigma_{xx} + s_{12}\sigma_{yy} + s_{13}\sigma_{zz} + s_{14}\tau_{yz} + s_{15}\tau_{zx} + s_{16}\tau_{xy}$$

and similar ones for the remaining five stress components, the simplification due, in particular to the elimination of the shear stresses, is clearly apparent. The assumption of isotropy has resulted in the reduction of the number of moduli to two, since, remembering that $s_{mn} = s_{nm}$,

$$s_{11} = s_{22} = s_{33} = 1/E \text{ and } s_{12} = s_{13} = s_{23} = -\nu/E$$

On rearranging equations 1.15 the stresses may be expressed in terms of the strains; the elastic constants are then found to be

$$c_{11} = c_{22} = c_{33} = E(1-\nu)/(1+\nu)(1-2\nu)$$

and

$$c_{12} = c_{13} = c_{23} = \nu E/(1+\nu)(1-2\nu)$$

The ratio $\nu E/(1+\nu)(1-2\nu)$ is sometimes denoted by λ, and

$$E/2(1+\nu) = G \tag{1.16}$$

Both λ and G are known as Lamé's constants; as we shall see later, G is the shear modulus.

The hydrostatic component of the strain, ϵ_h, accounts for changes in volume, but not in shape, and is therefore given by $\Delta V/V$, where ΔV is the change in volume of a cube of volume V under the action of the principal stresses σ_1, σ_2 and σ_3. If V is taken to be unity, ΔV is the hydrostatic strain. Now, taking $V = 1$:

$$1 + \Delta V = (1 + \epsilon_1)(1 + \epsilon_2)(1 + \epsilon_3)$$

On expanding the bracketed terms and neglecting terms of second and higher orders one obtains

$$\epsilon_h = \epsilon_1 + \epsilon_2 + \epsilon_3 \tag{1.17}$$

The relation

$$\sigma_h/\epsilon_h = K \tag{1.18}$$

with σ_h given by equation 1.5 defines the bulk modulus K, or its inverse, the compressibility. We note that for an incompressible material the volume strain must always be zero, so that then

$$\epsilon_1 + \epsilon_2 + \epsilon_3 = 0 \tag{1.19}$$

Since an equation analogous to 1.12 also holds for strains, one has for incompressible isotropic materials

$$\epsilon_{xx} + \epsilon_{yy} + \epsilon_{zz} = 0$$

independent of the choice of axes. In view of this relation the strain at a point of an incompressible elastic material is specified if five rather than six components of the strain tensor are known.

1.9 Energy Stored in Elastically Deformed Materials

The energy of elastic deformation may be regarded as potential energy, for it can be converted into kinetic energy, performing work as, for example, in a clockwork mechanism driven by a coiled metal spring. In the case of a wire or rod of cross-section A and length l_0 subjected to a tensile stress σ_1 the work stored is given by

$$w_1 = \int_{l_0}^{l} A \cdot \sigma \cdot \mathrm{d}l = A l_0 \int_{0}^{\epsilon_1} \sigma \, \mathrm{d}\epsilon$$

where $\epsilon_1 = \sigma_1/E$. Since Al_0 is the volume of the rod or wire, the energy stored per unit volume is given by

$$w_1 = E \int_{0}^{\epsilon_1} \epsilon \cdot \mathrm{d}\epsilon = \tfrac{1}{2} E \epsilon_1^2 = \tfrac{1}{2}\sigma_1 \cdot \epsilon_1 = \tfrac{1}{2}\sigma_1^2/E \qquad (1.20)$$

At a point in an elastic body where all the three principal stresses σ_1, σ_2 and σ_3 are non-zero the total elastic energy stored per unit volume is

$$w = w_1 + w_2 + w_3$$

and with the tensile strains given by equation 1.15 one obtains

$$w = \frac{1}{2E} \left[(\sigma_1^2 + \sigma_2^2 + \sigma_3^2) - 2\nu\,(\sigma_1\,\sigma_2 + \sigma_2\,\sigma_3 + \sigma_3\,\sigma_1)\right]$$

18

Now, w consists of two parts. One, w_s, arises from deviatoric strains and hence from changes of shape, while the second, w_h, is the hydrostatic component associated with changes in size. By analogy with equation 1.20 one finds

$$w_h = \tfrac{1}{2}\,\sigma_h \,.\, \epsilon_h \qquad (1.21)$$

with σ_h and ϵ_h given by equations 1.5 and 1.17 respectively. Thus one obtains

$$w_h = [(1 - 2\nu)/6E]\,(\sigma_1 + \sigma_2 + \sigma_3)^2 \qquad (1.22)$$

and, since $w_s = w - w_h$,

$$w_s = \frac{2(1+\nu)}{3E}\left[\left(\frac{\sigma_1 - \sigma_2}{2}\right)^2 + \left(\frac{\sigma_2 - \sigma_3}{2}\right)^2 + \left(\frac{\sigma_3 - \sigma_1}{2}\right)^2\right] \quad (1.23)$$

From equations 1.18 and 1.21

$$w_h = \frac{1}{2K}\left(\frac{\sigma_1 + \sigma_2 + \sigma_3}{3}\right)^2$$

and on comparing this with equation 1.22 the bulk modulus is obtained in terms of E and ν :

$$K = E/3(1-2\nu) \qquad (1.24)$$

For an incompressible material Poisson's ratio must therefore be equal to $\tfrac{1}{2}$.

Other relations between elastic constants of isotropic materials may be derived by analogous methods, applied however to w or w_s. For example, the elastic energy stored in a cube subjected to a simple shear, as in *Figure 1.7a*, is $\tfrac{1}{2}\tau_{zy}^2/G$, where the shear modulus is defined by $G = \tau_{zy}/\gamma_{zy}$. This energy can also be obtained by substituting for the principal stresses from equation 1.14 into the expression for the total stored energy derived above; the result is

$$w = \frac{1}{E}\,(1 + \nu)\tau_{zy}^2$$

19

Again, comparing both values, one derives the relation

$$G = E/2(1 + \nu) \qquad (1.25)$$

For incompressible materials

$$G = \tfrac{1}{3} E \qquad (1.26)$$

The last result may sometimes be of considerable use in practice, for example when E but not G are known, or vice versa. In the case of the more common metals, lead has a particularly low compressibility, with $\nu = 0 \cdot 44$. The elastic moduli, measured at room temperature, are $E = 1 \cdot 8 \times 10^5$ kg/cm² and $G = 0 \cdot 60 \times 10^5$ kg/cm²; the ratio E/G thus agrees with equation $1 \cdot 26$ reasonably well.

The decrease of the elastic constants with increasing temperature is comparatively small and, except close to the melting point, the variation can generally be represented with sufficient accuracy for most purposes by a linear relation. If, for example, the low-temperature value of the shear modulus is known (G_0), one may write

$$G\,(T) = G_0\left(1 - a\frac{T}{T_m}\right)$$

where T_m is the melting temperature, expressed in °K, and a is a constant equal to about $0 \cdot 2$.

1.10 Locating the Principal Stresses in Practice

In view of the extensive reference to principal axes which we have made in the preceding sections, we shall now give a simple example on their location in a practical case. The most important consideration is generally the fact that a surface not actually subjected to shearing forces must also be free from shear stresses. A principal stress therefore acts along the normal to that surface at the point considered. Knowledge of the direction of one more principal stress then suffices to locate the axes at the point. Symmetry or other geometrical features generally provide an indication of the second required direction.

In a thin-walled pressure vessel, which we take as our example, the internal pressure p does not result in a shear stress either in the shell or in the circular flat ends. Referring to *Figure 1.8*, we see that the principal stresses must in fact occur in axial, radial and circumferential directions. There is no tensile stress applied radially across the shell wall, and hence, in a thin-walled cylinder, $\sigma_3 = 0$. The stresses σ_1 and σ_2 are evaluated as follows.

Figure 1.8. Principal stresses in the shell of a thin-walled cylinder subjected to internal pressure, p

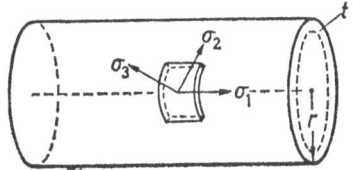

Firstly, the force on the cylindrical shell in the axial direction is $2\pi r t \sigma_1$, where r and t are the radius and wall thickness of the cylinder. This is in equilibrium with the force $\pi r^2 p$ acting on the flat base, so that

$$\sigma_1 = pr/2t$$

Secondly, by considering the equilibrium of forces on a horizontal, axial, section through the cylinder one has for unit length:

$$2\sigma_2 t = 2rp$$

so that

$$\sigma_2 = pr/t$$

The stress system is now fully specified.

MACROSCOPIC PLASTICITY

2.1 Yield Criteria for Isotropic Materials

In uniaxial tension the onset of plastic deformation is specified by the yield stress σ_y or, if the material has been work-hardened by previous plastic deformation, by the strain-dependent flow stress σ, for example that corresponding to the point P in *Figure 1.1*. Criteria for the onset of plasticity must necessarily be more complex in the case of triaxial stress systems, although they must of course include the uniaxial deformation as a special case.

Now, the change of shape of the material resulting from the plastic deformation should depend only on the deviatoric stresses, for even extremely high hydrostatic stresses are known not to lead to permanent, plastic, deformations of ductile materials, assuming them to be homogeneous. The yield criterion would therefore be expected to include only combinations of deviatoric stresses, and therefore to have the form

$$f(\sigma_1 - \sigma_2,\ \sigma_2 - \sigma_3,\ \sigma_3 - \sigma_1) = 0 \qquad (2.1)$$

in which the deviatoric stresses given by equation 1.3 have been combined in a simple, symmetric, manner, with the concomitant elimination of σ_h. The variables of the function f can be seen to be the principal shear stresses given by equation 1.9. Reversal of the signs of all the stresses should leave the function f invariant, for it is well known, for example, that a metal rod will yield plastically in tension at the same absolute value of the flow stress as in compression. This consideration suggests that f is a quadratic function of the principal shear stresses. It should also be symmetric with respect to the three principal stresses, for none of them is in any way preferred. A relation which satisfies these requirements is

$$(\sigma_1 - \sigma_2)^2 + (\sigma_2 - \sigma_3)^2 + (\sigma_3 - \sigma_1)^2 = 2\,\sigma^2 \quad (2.2)$$

where σ is a constant the significance of which has yet to be established; the factor 2 is used for convenience later. On comparing equations 2.2 and 1.23 the criterion is seen to imply that yielding will occur when the shear energy per unit volume attains a certain definite value characterised by the constant σ. If one considers the special case of a uniaxial tensile test, putting $\sigma_2 = \sigma_3 = 0$ into equation 2.2, one finds that σ is identical with the tensile flow stress of the material. This maximum-shear-energy criterion, originally proposed by Huber, is frequently also associated with the names of Hencky and von Mises. Its applicability to ductile solids, particularly to metals, has been confirmed by extensive tests.

A second criterion, often simpler to use in practice, is known as the maximum-shear-stress criterion; it is variously associated with the names of Coulomb, Tresca, Mohr and Guest. It states that yielding will occur when the largest principal shear stress attains a certain definite level, and may be written

$$\sigma_1 - \sigma_3 = \sigma; \qquad \sigma_1 > \sigma_2 > \sigma_3 \quad (2.3)$$

As before, σ is found to be the tensile flow stress of the material. The hydrostatic stress does not appear in the criterion, but the equation is not symmetric in the principal stresses, for the intermediate stress σ_2 is taken into account only through an inequality.

The two criteria can be compared by using the parameter

$$\mu = (2\sigma_2 - \sigma_1 - \sigma_3)/(\sigma_1 - \sigma_3)$$

which varies between $+1$ and -1 as σ_2 takes values between its upper and lower limits σ_1 and σ_3. Equation 2.2 can then be written

$$\sigma_1 - \sigma_3 = \sigma \ [2/(3 + \mu^2)^{\frac{1}{2}}]$$

showing that the value of $\sigma_1 - \sigma_3$ is at most by a factor $2/3^{\frac{1}{2}}$ greater than the corresponding value given by equation 2.3. The difference is therefore always less than about 16 per cent.

In the case of the thin-walled cylinder subjected to an internal pressure p, discussed in section 1.8, the maximum-shear-energy criterion shows that yielding will set in when p attains the value $\sigma t/3^{\frac{1}{2}}r$, while equation 2.3 yields $\sigma t/2r$; in this case $\mu = 0$. If it is desired to avoid plastic deformation of the cylinder then the maximum-shear-stress criterion will give a lower and hence 'safer' estimate of the permitted maximum pressure. In view of its simplicity it is frequently used in practice although equation 2.2 is generally found to agree somewhat better with experiment.

2.2 Yielding of Single Crystals

Plastic yielding in single crystals occurs by slip, sometimes also termed glide, in which thin lamellae of the crystal glide over neighbouring ones like playing cards in a packet. In metals the crystallographic slip planes which are preferred, i.e. along which slip takes place under the smallest shear stresses, are generally those most densely studded with atoms, and the directions of slip are along the shortest lattice spacings between crystallographically equivalent sites. In aluminium, copper, gold, lead, silver and many common metals having the face-centred cubic structure, slip takes place preferentially on the eight sets of ' octahedral ' planes, one of which is shown in *Figure 2.1a*. There are six possible slip directions in each plane, i.e. two along each face diagonal of the triangle shown. Altogether there are therefore 24 possible slip directions: two along each of the 12 face diagonals. In simple ionic crystals slip directions lie

24

generally along the joins of nearest neighbours of equally charged ions; *Figure 2.1b* shows a slip plane in a simple cubic crystal of the rocksalt type.

Glide occurs when the shear stress acting in the slip direction of a preferred system of glide planes attains a definite value, known as the critical resolved shear stress τ_{cr}. If, referring to *Figure 1.2*, a cylindrical crystal is subjected to uniaxial tension, and the angle between the normal of a glide plane and the tension axis is θ, while the glide

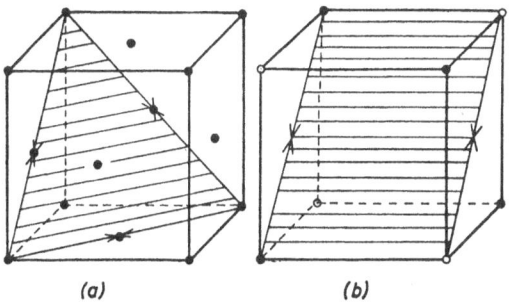

(a)   (b)

Figure 2.1. Preferred slip planes in: (a) face-centred cubic metals; (b) simple ionic crystals of the rocksalt type. Arrows indicate slip directions

direction makes an angle ϕ with the axis, then the component of the force along this direction is $F \cos \phi$. As the area of the glide plane is $A/\cos \theta$ the resolved shear stress in the glide direction is

$$\tau_{cr} = \sigma \cos \theta \cos \phi \qquad (2.4)$$

where $\sigma = F/A$. If several preferred glide systems exist, as for example in crystals of the type referred to in *Figure 2.1*, then on gradually increasing the tensile stress from zero a value of σ will be attained which will satisfy equation 2.4 for at least one of the slip planes. In fact the slip system with the highest value of the product $\cos \theta . \cos \phi$ will become operative first. There is extensive evidence in support of the critical shear stress law; its validity is readily checked experimentally by examining the constancy of this product with crystals variously oriented with respect to the tensile axis. Attempts have been made to show that equation 2.4 may be generalised so as to lead to yield criteria which are known to hold for polycrystalline aggregates, i.e. equations 2.2 or 2.3.

2.3 Equations of Plasticity

As we have already noted, a wire extended beyond the yield point acquires a permanent, plastic, deformation, and its length does not revert to its initial value if the stress is removed. Although the wire is now strained the applied stress is zero. Clearly, in view of this irreversibility of deformation, a linear relation between stress and strain, akin to Hooke's Law, cannot hold in the plastic range. However, we know that the previous stress would have to be applied, and in fact slightly exceeded, if we wished to extend the wire further by a small amount. One could therefore expect a relation between stresses and increments of plastic strain. In place of Hooke's Law, referred to principal axes, one may then write three differential equations.

$$\left. \begin{aligned} d\,\epsilon_{p1} &= dC.[\,\sigma_1 - \tfrac{1}{2}\,(\sigma_2 + \sigma_3)] \\ d\,\epsilon_{p2} &= dC.[\,\sigma_2 - \tfrac{1}{2}\,(\sigma_3 + \sigma_1)] \\ d\,\epsilon_{p3} &= dC.[\,\sigma_3 - \tfrac{1}{2}\,(\sigma_1 + \sigma_2)] \end{aligned} \right\} \qquad (2.5)$$

where C represents a yet undefined, strain-dependent, property of the material, and the subscript p indicates a plastic component. As irreversible deformations occur in most materials without significant volume changes, Poisson's ratio is taken equal to $\tfrac{1}{2}$, as required by equation 1.24.

Now we notice that

$$\sigma_1 - \tfrac{1}{2}\,(\sigma_2 + \sigma_3) = \tfrac{3}{2}\,(\sigma_1 - \sigma_h)$$

26

where σ_h is the hydrostatic stress given by equation 1.5. Analogous expressions hold for the other bracketed terms in equation 2.5. On writing σ_{d_1} for the deviatoric stress component $\sigma_1 - \sigma_h$, with a corresponding notation for $\sigma_2 - \sigma_h$ and $\sigma_3 - \sigma_h$, equation 2.5 can be recast into the form

$$\frac{\mathrm{d}\epsilon_{p_1}}{\sigma_{d_1}} = \frac{\mathrm{d}\epsilon_{p_2}}{\sigma_{d_2}} = \frac{\mathrm{d}\epsilon_{p_3}}{\sigma_{d_3}} = \tfrac{3}{2}\,\mathrm{d}C \qquad (2.6)$$

first derived by Prandtl and Reuss. The significance of $\mathrm{d}C$ can be established from the following considerations. The 'plastic' work per unit volume of solid due to simultaneous increments of the plastic strain is

$$\mathrm{d}w_p = \sigma_{d_1}\,\mathrm{d}\epsilon_{p_1} + \sigma_{d_2}\,\mathrm{d}\epsilon_{p_2} + \sigma_{d_3}\,\mathrm{d}\epsilon_{p_3}$$

which, in view of equation 2.6, may be written

$$\mathrm{d}w_p = \tfrac{3}{2}\,\mathrm{d}C\,(\sigma_{d_1}^2 + \sigma_{d_2}^2 + \sigma_{d_3}^2) \qquad (2.7)$$

It may readily be verified that equation 2.2 is equivalent to

$$\sigma_{d_1}^2 + \sigma_{d_2}^2 + \sigma_{d_3}^2 = \tfrac{2}{3}\,\sigma^2$$

so that

$$\mathrm{d}w_p = \sigma^2 \, . \, \mathrm{d}C \qquad (2.8)$$

in which the individual stress components do not appear explicitly. We may therefore consider the special case of a rod, or wire, extended plastically by uniaxial tension, the yield stress being σ. The increment of 'plastic' work is $\sigma.\mathrm{d}\epsilon_{p_1}$, so that equation 2.8 yields

$$\mathrm{d}\epsilon_{p_1} = \sigma \, . \, \mathrm{d}C$$

and on writing for the coefficient of work-hardening, given by the slope of the stress-strain curve at the flow stress σ,

$$H = \mathrm{d}\sigma/\mathrm{d}\epsilon_{p_1} \qquad (2.9)$$

one obtains

$$\mathrm{d}C = \mathrm{d}\sigma/H\sigma \qquad (2.10)$$

Thus if the yield or flow stress σ and the corresponding coefficient of work-hardening are known, it is possible to determine the strain increments resulting from a small increase of σ, as is apparent from equations 2.6 and 2.10. Integration of equation 2.6 is not however in general possible.

As a specific, simple, illustration we may consider the case where $H = \chi/\sigma$, χ being a constant, and $\sigma_1 \equiv \sigma$ is the only non-zero stress. Equations 2.6 and 2.10 then yield

$$d\epsilon_{p1} = (\sigma_1/\chi)d\sigma_1$$

and

$$\frac{d\epsilon_{p2}}{-\frac{1}{3}\sigma_1} = \frac{d\epsilon_{p3}}{-\frac{1}{3}\sigma_1} = \frac{d\epsilon_{p1}}{-\frac{1}{3}\sigma_1}$$

The solution

$$\sigma_1^2 = \chi\epsilon_{p1} ; \quad \epsilon_{p2} = \epsilon_{p3} = -\tfrac{1}{2}\epsilon_{p1}$$

shows that the wire, or rod, hardens 'parabolically', and also that the sum of the three plastic strain components is zero. The last result could of course have been expected because of the assumed incompressibility of the material (equation 1.19).

2.4 Approximate Methods and Applications

The mode of plastic deformation in wire drawing, deep drawing of metal sheet, extrusion, rolling and other shaping processes is generally too complex to yield to rigorous analysis, but results useful in practice may still be obtained if reasonable simplifying assumptions are made. The large deformations which are as a rule involved also call for a redefinition of strain; clearly it would not be convenient or physically meaningful to express, let us say, the plastic strain of a long thin wire in terms of the initial dimensions of the relatively small metal billet from which it was drawn.

Again consider a rod of initial cross-section A_1 and length l_1 ; the material is assumed incompressible. As its volume is therefore constant one has

$$A_1 l_1 = A_2 l_2 = Al = V = \text{constant} \qquad (2.11)$$

where A is the cross-section when the rod has been extended to length l, and A_1 and A_2 refer to the initial and final states respectively. The flow stress σ at any stage of the deformation will depend upon the plastic strain, as is clear from *Figure 1.1*. The 'plastic' work expended in the deformation is

$$W_p = \int_{l_1}^{l_2} \sigma(l) \,.\, A \,.\, \mathrm{d}l = V \int_{l_1}^{l_2} \sigma(l) \frac{\mathrm{d}l}{l} \qquad (2.12)$$

the second integral being obtained by the use of equation 2.11. Hence the work per unit volume, W_p/V, is

$$w_p = \int_{l_1}^{l_2} \sigma\,(l) \,.\, \mathrm{d}\,(\ln l) \qquad (2.13)$$

On comparing equations 2.12 and 2.13 with corresponding expressions for the work done in elastic deformations, given in section 1.9, one finds that the tensile strain is now replaced by the natural logarithm of the instantaneous length of the rod. This so-called 'natural strain', sustained by the rod on being extended from length l_1 to l_2, is $\ln l_2 - \ln l_1$, or $\ln (l_2/l_1)$. This may also be written $\ln[1 + (\Delta l/l_1)]$ where $\Delta l = l_2 - l_1$. If $\Delta l \ll l_1$ the logarithm may be expanded as a power series, and terms of second and higher degrees in $\Delta l/l_1$ may be neglected. The natural strain is then found to be equal to the conventional strain $\Delta l/l_1$.

Equation 2.13 also implies that the area under the stress versus natural strain curve, bounded by two values of the strain, is numerically equal to the work per unit volume expended in straining the rod from length l_1 to l_2. It is sometimes convenient to replace the actual stress–strain curve by one representing an 'ideally' plastic material, as shown in *Figure 2.2*.

If the level of the yield stress of the ideally plastic body is judiciously chosen, problems in plasticity may become mathematically simplified without undue sacrifice of pre-

29

cision; this applies particularly at strains where the slope of the work-hardening curve is relatively small.

Taking σ_0 to be the yield stress of the ideally plastic solid used to replace the actual one (*Figure 2.2*), equation 2.13 reduces to the simple relation

$$w_p = \sigma_0 \ln (l_2/l_1) = \sigma_0 \ln (A_1/A_2) \qquad (2.14)$$

The form involving the areas follows from the constancy of volume as expressed by equation 2.11.

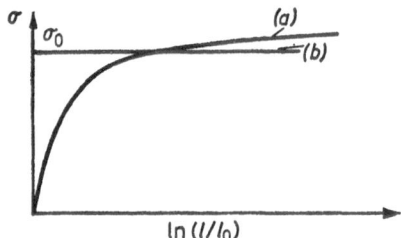

Figure 2.2. (a) The actual natural stress-strain curve; (b) its approximation by an ideally plastic body

The rod could also have been extended by extrusion, drawing through a die, or by rolling. In such operations the useful work necessary to impart the required change of shape is always accompanied by some wasted 'plastic' work, so that the efficiency of the process, ψ, defined as the ratio of useful to total work expended, is less than 1. The origin of some of the wastage of work can be seen by considering extrusion. As is apparent from *Figure 2.3*, a short horizontal element of the material in the billet container is first bent into a convex shape and subsequently straightened as it passes through the die or 'virtual die' formed by the static 'dead' material at the container wall around the exit orifice. The bending requires expenditure of 'plastic' work without contributing to the required change of shape.

Thus apart from frictional effects at the billet–container interface, which can often be made relatively unimportant by appropriate lubrication, allowance has to be made for this. Equation 2.14 has then to be modified to

$$w_p = (\sigma_0/\psi) \ln(A_1/A_2) \qquad (2.15)$$

Values of ψ generally lie between $0\cdot5$ and $0\cdot8$ for most simple forming processes.

The force F_1 which has to be applied to the ram (*Figure 2.3*) to effect extrusion can be evaluated as follows. Neglecting billet–container friction, and assuming an ideally plastic material we see that the work $F_1 x$ done in displacing the

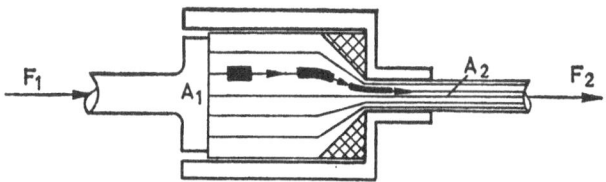

Figure 2.3. Extrusion v fo rod of cross-section A_2 from a billet of cross-section A_1. Reversed bending of an element indicates expenditure of 'useless' work

ram a distance x towards the die leads to the extrusion of a volume $A_1 x$ of material, and consequently to an expenditure of work $A_1 x w_p$. On equating this to $F_1 x$, and using equation 2.15, one obtains

$$F_1 = A_1 (\sigma_0/\psi) \ln(A_1/A_2) \qquad (2.16)$$

If the material is drawn through a die by a force F_2 instead then similar reasoning shows that

$$F_2 = (A_2/A_1)F_1$$

where F_1 is given by equation 2.16. The stress on the material due to the applied drawing force F_2 is equal to about $(\sigma_0/\psi) \ln (A_1/A_2)$, and this must be less than σ_0 if the wire, or rod, is not to break. Hence an upper limit is set to the possible reduction in area per pass.

The preceding principles are readily adapted to other shaping processes, such as draw rolling—in which metal sheet is pulled by a coiler drum through free-wheeling rolls —or to deep drawing. In the former case, *Figure 2.4*, the

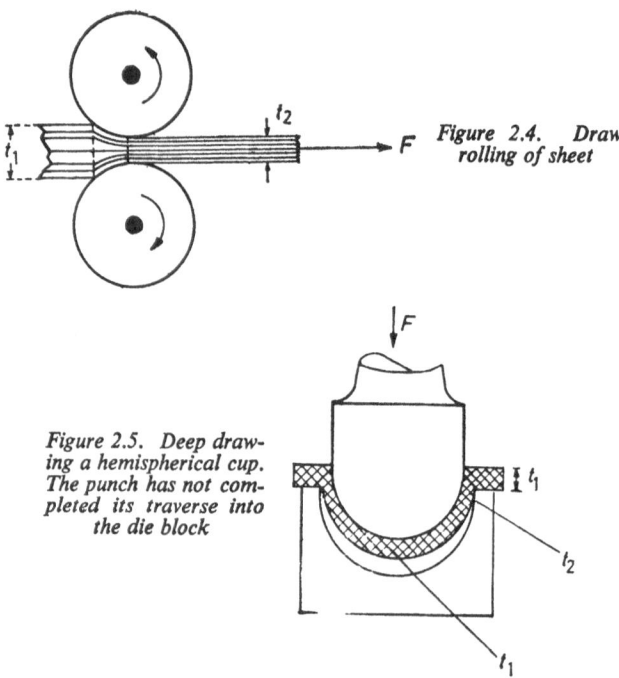

Figure 2.4. Draw rolling of sheet

Figure 2.5. Deep drawing a hemispherical cup. The punch has not completed its traverse into the die block

analogy with extrusion is obvious, while in deep drawing, *Figure 2.5*, it may be discovered as follows.

We consider a hemispherical metal cup drawn from a circular blank of equal diameter. In the course of being drawn it assumes intermediate shapes, as shown shaded in the figure. The thickness at the rim is equal to the final width, t_2 but at the centre of the base the thickness is still the same as that of the blank. The effective die may therefore be considered to extend at any time during the drawing process over the parts of the cup where the thickness lies between the initial and final values, t_1 and t_2. The area of the part of the blank converted into the cup is equal to $\frac{1}{2}$ that of the cup, so that

$$w_p = (\sigma_0/\psi)\ \ln 2$$

As the punch travels into the die-block a total distance approximately equal to the radius r of the cup one has, assuming uniformity of deformation,

$$F\ .\ r = V\ .\ w_p$$

where V is the volume of a circular area of the blank of radius r, i.e. $\pi\ r^2 t_1$. Thus

$$F = (\pi\ r\ t_1\ \sigma_0/\psi)\ \ln 2$$

In hot shaping operations, when the temperature of the material is in excess of about $0\cdot4\ T_m$ (°K), where T_m is the melting temperature, the flow stress generally increases significantly with increasing strain rate, an effect which will be discussed later in connection with high-temperature creep. A knowledge of the strain rate may therefore be important in the determination of ram pressures, press capacities, and so on. We shall indicate here how the strain rate may be estimated in the case of extrusion; similar methods appropriate to other processes suggest themselves readily.

In the corner of the container, *Figure 2.3*, an accumulation of static ' dead ' material forms a virtual die in the shape of a truncated, approximately rectangular cone; the planes of maximum shear stress forming the cone surface in general make an angle of about 45° with the billet axis. Deformation of the material occurs mainly within this die. Now the passage of a small volume of material, initially at the die mouth, through the die will take place over an interval of time given by the ratio of die volume to volume of material extruded per unit of time. On taking the die height to be approximately $\frac{1}{2}D$, where D is the billet diameter, the time of passage is found to be

$$t = (\pi D^3/24)/(\pi D^2 v/4) = D/6v$$

where v is the ram velocity. The strain rate is therefore

$$\frac{\mathrm{d}\epsilon}{\mathrm{d}t} = \frac{12v}{D}\ln\frac{D}{d}$$

since $\ln(A_1/A_2) = 2\ln(D/d)$, d being the diameter of the extruded rod.

2.5 The Stress–Strain Curve

The mechanical behaviour of materials is generally investigated by tests each of which is characterised by the use of a specific stress system, such as tension, compression, torsion, bending, or a combination of these, as well as by the time rate of application of the stresses and the total time under load. Thus if the time rate is zero, the material being subjected to constant stresses for long periods, we have conditions used in studies of creep; bending with impact loading occurs in studies of brittleness by the notched-bar method, while prolonged application of periodically varying stresses is employed in the investigation of fatigue. At present we shall confine our attention to the tensile test, as an example of a widely used ' static ' method of studying the plastic behaviour of materials. The principles involved in the measurement of hardness, impact strength and toughness, in all of which loading is relatively rapid, i.e. ' dynamic', we shall briefly examine later.

In a tensile test a specimen in the form of a rod, wire or strip, is subjected to gradually increasing loads and the relation between the tensile force and the extension is recorded. The machine may extend the test piece at a constant velocity, and the strain rate is then roughly constant in the course of extension; alternatively a nearly constant rate of stressing may be obtained by pouring liquid or granular material into a loading pan or bucket suspended from the specimen. In the first case the machine is said to be 'hard', the second arrangement is termed 'soft.' Few machines are either purely hard or soft, for it is in practice rather difficult to achieve strict control over the rates of strain or stress.

The well-known 'yield-drop' of the load which is observed with certain alloys, such as mild steel, and brasses

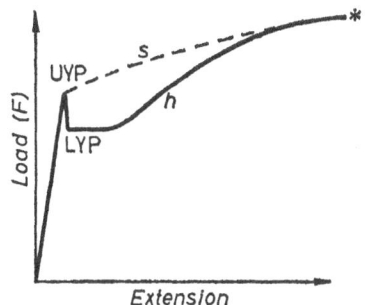

Figure 2.6. Load–extension curves for a material with a pronounced yield effect, obtained on a hard (h) and soft (s) tensile testing machine

containing 20–30 per cent by weight of zinc, indicated in *Figure 2.6*, could not occur with a soft machine, for a drop in the applied load is precluded by the constancy and positive value of the loading rate. The characteristics of the machine may therefore influence the load–elongation curve appreciably under certain circumstances.

A typical load–elongation curve, such as may be obtained with a great variety of materials, is shown in *Figure 2.7*. Provided the material is ductile, so that fracture, indicated by a star, does not occur at too small a strain, the curve is found to become convex, with a definite maximum at a

35

load F_{max}. This maximum load, divided by the cross-section of the undeformed test piece, A_0, is referred to as the ultimate tensile strength, σ_{ult}, often also denoted by UTS.

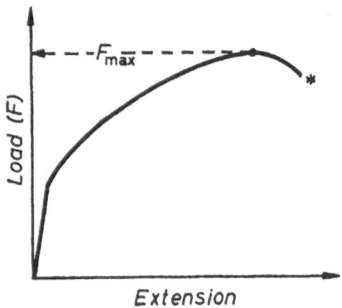

Figure 2.7. Typical load–extension curve of a ductile material, with a maximum attained before fracture

Its occurrence is a consequence of the method of testing; as we shall show it does not correspond to any sudden changes in the physical properties of the material, although it may provide a practical measure of the load-bearing capacity of a metal rod, for example.

By definition we have

$$\sigma_{\text{ult}} = F_{\text{max}}/A_0 \qquad (2.17)$$

while the corresponding true stress, referred to the actual cross-section A_{ult} is

$$\sigma_{\text{max}} = F_{\text{max}}/A_{\text{ult}} \qquad (2.18)$$

If the true stress corresponding to an instantaneous rod cross-section A is σ, then

$$F = A\sigma$$

The force will attain its maximum F_{max} when

$$\mathrm{d}(A\sigma) = 0 \qquad (2.19)$$

i.e. when

$$-\,\mathrm{d}A/A = \mathrm{d}\sigma/\sigma$$

36

As we are assuming the material to be incompressible it follows from equation 2.11 that $-\,\mathrm{d}A/A = \mathrm{d}l/l$, where l is the instantaneous length of the specimen. We may therefore write

$$\mathrm{d}\sigma/\sigma = (\mathrm{d}l/l_0)\,(l_0/l)$$

or

$$\mathrm{d}\sigma/\sigma = \frac{\mathrm{d}\,(l - l_0)}{l_0} \cdot \frac{l_0}{l_0 + \Delta l}$$

which yields

$$\frac{\mathrm{d}\sigma}{\mathrm{d}\epsilon} = \frac{\sigma}{1 + \epsilon} \tag{2.20}$$

As equation 2.19 was used in deriving this result, $\mathrm{d}\sigma/\mathrm{d}\epsilon$ is the slope of the stress–strain curve where, by equation 2.18, the stress is equal to σ_{max}. Hence σ_{max} may be found on the

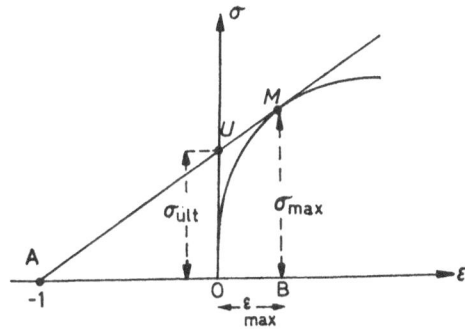

Figure 2.8. Determination of the ultimate tensile strength from the true stress-strain curve

true stress–strain curve as the point M of contact of the tangent to it drawn from the point $\epsilon = -\,1$ on the strain axis. This is apparent from *Figure 2.8*. The magnitude of the ultimate stress is then obtained from the relation

$$\sigma_{ult}/\sigma_{max} = A_{ult}/A_0 = l_0/l = 1/1 + \epsilon_{max}$$

37

which follows from equations 2.17 and 2.18. The geometrical significance of this equation can be seen by considering the similar triangles ABM and AOU in *Figure 2.8*; clearly σ_{ult} is represented by the intercept OU which the tangent makes on the stress axis. The existence of a maximum in the force–elongation curve is therefore seen to result from the failure to allow for the continuous decrease of the test piece cross-section in the course of deformation in that representation.

In practice the UTS provides an indication of the stress at which local constrictions, also termed 'necks,' begin to form in the test piece. Deformation then tends to occur preferentially at these points, as the tensile stress is there above the average for the specimen. The material may tend to draw down into a double cone; in ductile metals a characteristic cup-and-cone fracture then often results. The strain at fracture or the reduction in cross-section at the neck at which failure occurred are used as criteria of ductility in practice.

2.6 Hardness and Toughness

The resistance of plastic materials to deformation is frequently assessed by their hardness, which is also an important, though by no means sole, parameter determining their resistance to abrasion and wear. Most measurements of hardness are based on the determination of the area of an indentation made by a pyramidal diamond (Vickers, Knoop) or spherical metal indentor (Brinell, Rockwell) under a definite load, which is allowed to act for a specified time ranging from a fraction of a second to a few seconds.

We shall discuss the Brinell test as a typical example of the use of a spherical indentor, in this case a hardened steel sphere, and the Vickers test in which a diamond pyramid with a square base is used.

The Brinell hardness number (BHN) is defined as the stress W/A kg/mm^2, where W is the load applied to the

38

indentor, and A the surface area of the spherical cap forming the indentation. If D and d are the indentor and indentation diameters respectively, one has

$$\text{BHN} = W/(\tfrac{1}{2}\pi D^2)\left\{ 1 - [1 - (d/D)^2]^{\frac{1}{2}} \right\}$$

or

$$\text{BHN} = [W/(\tfrac{1}{4}\pi d^2)]\, \text{f}(d/D) \qquad (2.21)$$

where

$$\text{f}(d/D) = \tfrac{1}{2}\, (d/D)^2/\left\{ 1 - [1 - (d/D)^2]^{\frac{1}{2}} \right\}$$

Now, if the indentation is small, so that $d \ll D$, then expansion of the term in square brackets and neglect of second and higher order terms in d/D yields $\text{f}(d/D) = 1$, while with large indentations, when $d/D \approx 1$, $\text{f}(d/D) \approx \tfrac{1}{2}$. It follows that consistent results irrespective of indentor diameter can be obtained only if the ratio d/D is maintained constant, and preferably small, the latter requirement being desirable to facilitate ready conversion of the BHN to other hardness numbers, as will become apparent below. In practice the indentor is kept under load for about 20 seconds, and the diameter of the identation is then measured by means of a low-power microscope. The result is used to obtain the BHN directly from tables.

In the Vickers 'pyramid hardness test' a hydraulic mechanism applies a load W for a fraction of a second to a diamond indentor initially arranged so as to touch the surface of the material with its tip. The pyramid has a square base, and opposite faces make an angle of 136° with one another. The use of a large angle helps to minimise friction effects.

If one assumes that a uniform normal pressure p acts on the indentor, then by considering a face, such as AOB in *Figure 2.9*, the vertical component of force opposing the load on this face is seen to consist of contributions $2x\mathrm{d}l \cdot p \cdot \sin \alpha$ and $2x\mathrm{d}l.\mu p.\cos \alpha$, where μ is the coefficient of friction at the diamond–material interface, and $\mathrm{d}l$ the width

39

of the elementary strip a distance x along the face from the apex. In equilibrium

$$W = 4p \int_{x=0}^{\frac{1}{2}a} (\sin \alpha + \mu \cos \alpha) \cdot 2x \cdot dl$$

or

$$W = p (1 + \mu \cot \alpha) \int_{x=0}^{\frac{1}{2}a} 8x \cdot dx$$

The integral yields the projected area of indentation, a^2, and the VHN, given by W/a^2, is then

$$\text{VHN} = p (1 + \mu \cot \alpha) \qquad (2.22)$$

With the large values of the cone semi-angle used the friction term $\mu \cot \alpha$ is generally negligible compared with 1, and the VHN is then equal to p. In practice the indentation diagonal is measured, and the VHN is then read directly from tables. It can now be seen that the BHN and VHN should be equal for a given material, provided f(d/D) in equation 2.21 is close to 1, i.e. $d/D \ll 1$. Similarity requirements such as relate to f(d/D) do not arise with pyramidal indentors.

Most materials become harder and less ductile as the temperature is lowered; well known examples include sealing wax, rubber and metals. A piece of rubber submerged in liquid nitrogen (77°K) and then withdrawn can be shattered with a hammer; an abrupt loss of ductility also occurs in most common types of steel, generally in the range 0 to -40°C. Resistance to impact loading of embrittled materials is low; fracture is readily initiated if notches, scratches and other stress-raising defects occur in the material. Sometimes notch sensitivity of this type is useful, for example on cutting sheet glass with diamond tipped tools.

An embrittled material will fracture without significant expenditure of work on plastic deformation; in fact the work expended in fracturing standard test pieces provided with rectangular notches of standard depth and root radii is used in the Charpy and Izod tests for comparing toughness or, as it is sometimes called, 'notch brittleness' of materials.

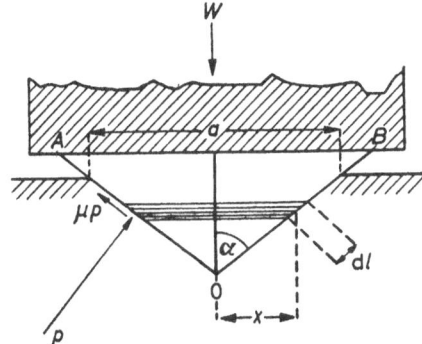

Figure 2.9. Indentation of an ideally plastic material by a diamond pyramid indentor. AOB is one of the four indentor faces

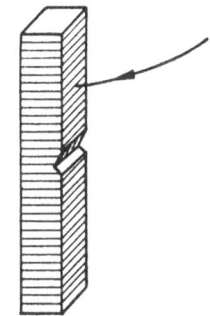

Figure 2.10. A 'Charpy' V-notch test piece for notch impact strength determinations. Depth of notch is 2 mm, notch root radius is 0·25 mm

Figure 2.10 shows a Charpy V-notch specimen used for impact testing. The sides of the square base measure 1 cm, and the notch, situated at the centre of the 5·5 cm long specimen, is 2 mm deep. The notch root radius is 0·25 mm.

The bottom part of the specimen is securely clamped and the top is struck from the notch side by a pendulum hammer, which loses energy in fracturing the specimen. The difference between the down and up-swing angles of the pendulum, read from a scale, yields the energy absorbed in fracturing the specimen. It is expressed in kg m and provides a measure of the notch impact strength, useful in particular in investigating the effect of temperature changes, heat treatments and alloying, on the toughness of the material.

3

PLASTICITY AND STRENGTH OF CRYSTALS

3.1 The Theoretical Strength of Crystals

The shear stress at which plastic deformation should begin to take place in an ideally perfect crystal was first calculated by Ya. Frenkel in 1926. The method is straightforward. Referring to *Figure 3.1*, which represents two adjacent planes of atoms of a crystal of simple structure, we see that

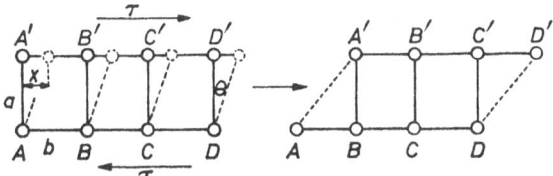

Figure 3.1. Slip in an ideally perfect crystal

if the upper one were displaced relative to the lower one in the slip direction b so that the atoms A', B', C', etc. came to lie over B, C, D, etc. respectively, slip would have taken place. The crystal would remain crystallographically perfect as the atoms have moved to equivalent lattice sites. The shear stress necessary to produce this permanent plastic deformation is the critical shear stress τ_c of the ideally perfect crystal.

To obtain its value we first consider the form of the stress-displacement curve (*Figure 3.2*), measuring displacements x indicated in *Figure 3.1*. Now $\tau = 0$ when $x = 0$ and $x = b$; the crystal is in stable equilibrium in both cases.

When $x = \tfrac{1}{2}b$ the atoms in the upper layer can be seen to be in a position of unstable equilibrium; the force in the slip direction is zero, for each atom is equally attracted to the right and to the left. Consequently $\tau = 0$ also at $x = \tfrac{1}{2}b$. Further, we know from the elastic behaviour of crystals that the shear stress increases linearly with the strain within

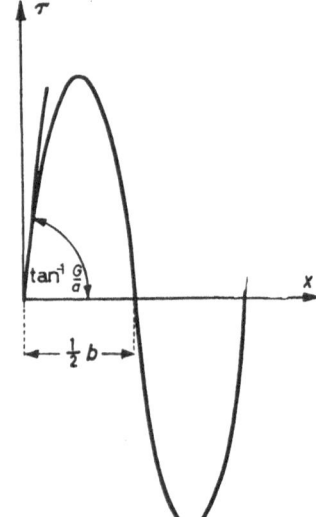

Figure 3.2. The harmonic stress versus displacement law for slip in an ideally perfect crystal

the elastic range. The slope of the shear stress versus displacement curve at $x = 0$ and $x = b$ must therefore comply with Hooke's Law

$$\tau/\gamma = G \qquad (3.1)$$

where

$$\gamma = x/a$$

Thus

$$\mathrm{d}\tau/\mathrm{d}x = G/a, \qquad x \ll a \qquad (3.2)$$

44

Now $\tau(x)$ must be a periodic function with wavelength b, and we therefore take the relation

$$\tau = K \,.\, \sin\,(2\pi x/b) \qquad (3.3)$$

shown in *Figure 3.2*. Other harmonic functions satisfying the boundary conditions at $x = 0$, $\tfrac{1}{2}b$ and b could have been chosen, without however affecting the result significantly. As equation 3.3 must reduce to equation 3.1 for small strains, one finds that then

$$\tau = Gx/a = K(2\pi x/b)$$

so that

$$K = Ga/2\pi b \qquad (3.4)$$

The highest shear stress occurs when

$$2\pi x/b = \tfrac{1}{2}\pi$$

i.e. at $x = \tfrac{1}{4}b$, and is numerically equal to K. In practice a and b will not differ from one another appreciably, and one therefore obtains the result that the critical shear stress of an ideally perfect crystal is

$$\tau_c \approx G/10 \qquad (3.5)$$

The maximum ' elastic ' shear strain also occurs when x is $\tfrac{1}{4}b$, and is therefore equal to $\tfrac{1}{4}b/a$ which, with $a \approx b$, yields

$$\gamma_c \approx 25 \text{ per cent} \qquad (3.6)$$

Crystals of the high degree of perfection necessary to check this theory can be obtained readily only in the form of microscopically thin ' whiskers'. They have nevertheless permitted sufficient experimentation to establish the correctness of the foregoing results. However, ordinarily the critical shear stress of, say, pure copper crystals is about 20 kg/cm^2, while $G/10$ (equation 3.5) yields about 40,000 kg/cm^2; the strength of the ideal crystal is therefore 2,000 times higher than that of a real one. Similar ratios are obtained with other metallic and non-metallic crystals. The disagreement persists with polycrystalline materials although the ratio of ideal strength to true shear strength is then somewhat less than the corresponding fraction for single crystals.

The origin of this extreme weakness of real crystals, compared with ideal ones, was the subject of extensive speculation over a period of several years, but remained obscure until 1934–35 when G. I. Taylor, E. Orowán, M. Polányi, as well as Ya. Frenkel jointly with T. Kontorova, in independent papers laid the foundations for a rational theory of crystal plasticity. The common feature of these developments was the hypothesis that transport of matter, giving rise to plasticity, occurred through the agency of specific defects of the crystal lattice termed ' dislocations '. Direct evidence of the existence and behaviour of dislocations was not obtained until two decades later, when etch-pit studies and electron transmission microscopy methods confirmed the principal characteristics of dislocations which had until then been inferred mainly from theoretical considerations. In the following section we shall outline the features of the Theory of Dislocations essential to an understanding of crystal plasticity.

3.2 Edge and Screw Dislocations

An interesting analogy illustrating the role of dislocations in slip, suggested by N. F. Mott, is based on the introduc-

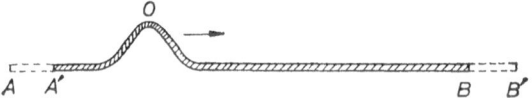

Figure 3.3. Illustration of the function of an edge dislocation. Movement of the ruck, O, transports matter from left to right

tion of a ruck, *O*, into a long runner carpet which it is desired to shift from its initial position *AB* to a new one *A' B'* (*Figure 3.3*).

The ruck, which carries the excess material originally at AA', is readily impelled by a small force to move in the direction indicated, leading to the displacement of the entire carpet which, by pulling at B, would have been hard to move bodily. A similar ruck may be introduced into an elastic cylinder by slitting it axially up to some line OO' (*Figure 3.4*) and then re-sealing it with the adjacent faces out of alignment by an amount AA'. The line OO' at which the ruck is located represents an edge dislocation in the elastic cylinder.

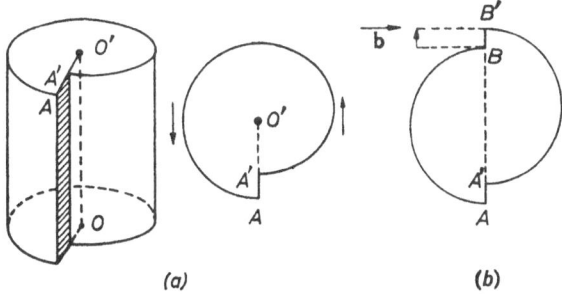

Figure 3.4. An edge dislocation of Burgers vector **b**: (a) *inside an elastic cylinder;* (b) *slip due to its passage through the cylinder*

If the cut were extended so that the edge OO' moved further into the material the dislocation could move right through the cylinder under the action of a shear stress τ acting in the slip plane $AO'O$, and both halves would become displaced with respect to one another by an amount AA, as is apparent from *Figure 3.4b*. The cylinder is therefore sheared as a result of the passage of the dislocation through it. AA' (or BB') is known as the Burgers vector.

In a homogeneous elastic material, such as rubber, the length AA' of the slip vector may be chosen arbitrarily, but in crystalline substances the directions and lengths must be those of a definite lattice spacing, as was already pointed out in section 2.2. *Figure 3.5* shows the passage of an edge dislocation lying at right-angles to the plane of the paper through a crystal.

Remember that the crystal is three-dimensional; *Figure 3.5* shows that the dislocation may be considered to result from the insertion of an extra plane of atoms up to half way down the crystal, its edge terminating in the slip plane along which shearing occurs. In the figure the extra half plane and the slip plane immediately adjacent to the edge,

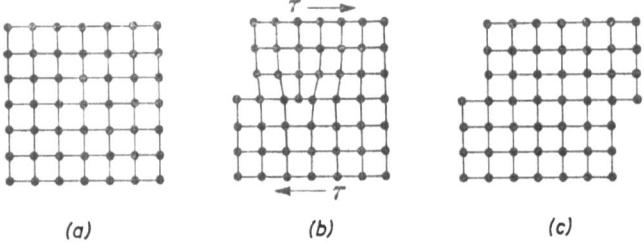

(a)	(b)	(c)

Figure 3.5. Passage of an edge dislocation through a crystal: (a) perfect crystal; (b) dislocated crystal; (c) sheared, dislocation-free crystal

i.e. at the centre of the dislocation, are seen to form an inverted letter T, and the symbol $\perp$ is generally used to represent a positive edge dislocation. The sign, according to convention, is negative if the half-plane is inserted into the crystal from below; the appropriate symbol is then $\top$.

Atoms adjacent to the edge, such as those along lattice lines below the slip plane are held in position relatively weakly and, in view of the looseness of the lattice in their neighbourhood, they will readily move into alignment below the centre of the dislocation. Now if, under an applied shear stress, the nearest line is drawn below the centre of the edge, then lower lines will follow, and the half-plane is converted to a normal, complete, plane of atoms. Although the dislocation thereby disappears from its initial position, transfer of the lower lines of atoms

48

has now left a half-plane of atoms adjacent to its previous location; the process is therefore the mechanism by means of which the dislocation propagates and transports material. By contrast with the Frenkel process, slip now propagates through the crystal in stages; it does not occur by simultaneous displacement of all atoms on the operative slip plane.

In the case of a straight edge dislocation normal to the crystal face (*Figure 3.5b*) the stress necessary to induce its migration, and hence slip, known as the Peierls–Nabarro stress, can be shown to be given by

$$\tau_{PN} = [2G/(1 - \nu)] \exp(-4\pi\zeta/b) \qquad (3.7)$$

where ζ, the half-width of the dislocation, is equal to the distance from the centre of the dislocation, measured along the slip plane, up to the point where the displacement of the atoms falls to one half of its maximum value. In the simple cubic lattice shown in *Figure 3.5*, $\zeta = a/2(1 - \nu)$ where a is the lattice spacing. As this is equal to **b**, the Burgers vector, then with $\nu = \frac{1}{3}$ the exponent in equation 3.7 is equal to about -9; for face-centred cubic metals one obtains approximately the same value. For copper, in particular, τ_{PN} is found to be close to 150 kg/cm^2, which is still several times higher than the experimentally determined critical shear stress of well annealed copper crystals. Refinements of the Peierls–Nabarro treatment have been proposed, and these lead to a significant reduction of the calculated critical shear stress, although even then the discrepancy is not satisfactorily resolved. However, direct observation of dislocations by transmission electron microscopy has shown that, contrary to the assumptions made in the Peierls–Nabarro theory, dislocations are not confined to specific lattice directions, but generally lie across rather than in the potential valleys in the crystal. A stress appreciably less than τ_{PN} as given by equation 3.7 should suffice to move such dislocations.

49

It should be noted that the atoms at the edge of the extra plane shown in *Figure 3.5b* have fewer nearest neighbours than other atoms. In covalent crystals, such as germanium or silicon, unsaturated 'dangling' bonds must then exist at the edge, and the dislocations will therefore

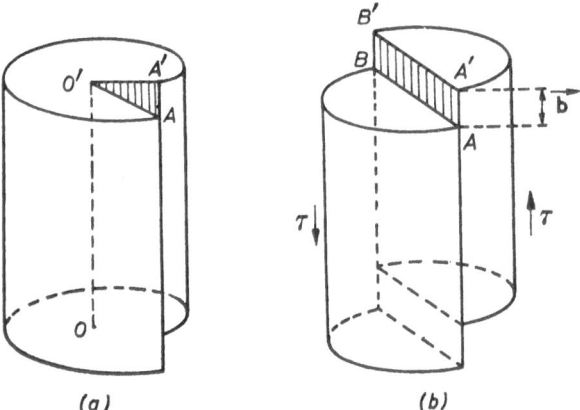

Figure 3.6. A screw dislocation of Burgers vector **b**: (a) *inside an elastic cylinder;* (b) *slip due to its passage through the cylinder*

be electrically charged. Impurity charge carriers of opposite sign will as a result be attracted to the dislocation and, if they can diffuse through the lattice, they will migrate towards the dislocation and attach themselves to the free bond.

Another type of dislocation of considerable importance in the growth of crystals from the vapour phase and in plasticity is known as the 'screw' dislocation. In order to examine the geometry of such a dislocation we again cut an elastic cylinder as before, but we now displace the adjacent faces of the cut vertically with respect to one another, as shown in *Figure 3.6*. If the dislocation *OO'* is forced to pass through the cylinder from front to back, by the applica-

tion of a shear stress as indicated, one half will be displaced by an amount equal to the Burgers vector along AA' with respect to the adjacent one. If a complete circle is drawn around the dislocation line OO', going clockwise along the rim and starting at A, it will terminate at A', a distance b above the starting point. The path is in fact helical rather than circular, as on a screw thread; the name of the dislocation refers to this property.

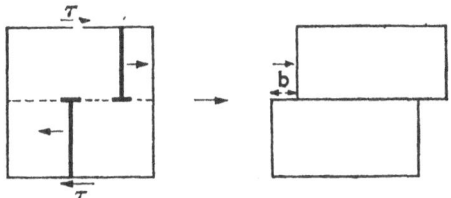

Figure 3.7. Shear of a crystal cube due to the passage of two dislocations of opposite signs

Again, in a crystalline material the Burgers vector AA' must be equal to a definite lattice spacing.

It should be noted that the Burgers vector lies in the direction of the dislocation line OO', while in the case of an edge dislocation (*Figure 3.4*) it is perpendicular to the dislocation line. This distinction will be of particular significance in the consideration of the spread of slip by a dislocation loop, which will be considered below. As a preliminary we shall examine the behaviour of two edge dislocations of the same Burgers vector lying on the same slip plane, when a shear stress is applied to the latter. As can be seen from *Figure 3.7*, if the shear stress is applied as indicated, the upper dislocation will move to the right and the lower one to the left, and the crystal will be sheared as shown; the effect is the same as would be obtained by the passage of a single dislocation right through the crystal. If however the sense of the shear stress were reversed the dislocations

51

would move towards one another and, on meeting, a complete unfaulted plane of atoms would form, with the consequent disappearance of both dislocations. This phenomenon has its parallel with screw dislocations, and may be generalised in the rule that coalescence of two dislocations of the same type but opposite signs results in their mutual annihilation.

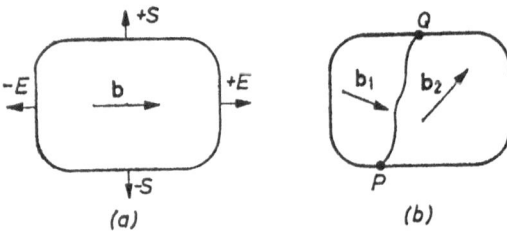

Figure 3.8. Dislocation loops (slip plane parallel to the surface of the paper): (a) all parts of the loop have the same Burgers vector, and the loop expands under an appropriate shear stress as indicated by the arrows at the edges; (b) a loop with different Burgers vectors in parts separated by the line PQ

Now, returning to the analogy of the runner carpet, we see that the ruck representing the dislocation line cannot terminate inside the carpet; it must stretch right across its entire width. In fact, the ruck may be regarded as the boundary line separating the slipped from the yet ' unslipped ' part of the carpet. Similarly, a dislocation cannot terminate within a crystal. It must either terminate at the crystal walls, or it must exist in the form of a closed loop enclosing a slipped area, as shown in *Figure 3.8a*.

It is clear from the discussion of the characteristics of edge and screw dislocations that if the positive edge dislocation moves as indicated under a shear stress in the slip plane, the negative edge dislocation $-E$, as well as the screws $+S$ and $-S$ will also move outwards, away from the centre of

52

the loop, thus expanding the slipped area. That the Burgers vector of a closed loop must in fact be the same over its entire length can be shown by considering a loop (*Figure 3.8b*) divided by a line PQ into two parts with Burgers vectors $\mathbf{b}_1$ and $\mathbf{b}_2$ respectively. This implies that the area left of PQ has slipped by an amount differing from that by which the part to the right of PQ has slipped. The difference is in fact $\mathbf{b}_1 - \mathbf{b}_2$, so that PQ must be a dislocation having this Burgers vector. It follows that a dislocation loop, free from network nodes such as PQ, has the same Burgers vector throughout.

In general the corners of isolated loops will tend to be rounded, consisting of dislocations of the ' mixed ' type, which have partly edge and partly screw characteristics, behaving essentially as if they consisted of small segments of alternate edge and screw type. The question of the curvature of dislocations will be discussed in some detail later.

3.3 Stresses due to Dislocations

As can be seen from *Figure 3.5*, the crystal is compressed above the slip plane near the centre of a positive edge dislocation; below the slip plane it is in tension. The converse is of course true for a negative edge dislocation. Consequently the elastic, potential, energy stored in the lattice around two edge dislocations of the same Burgers vector but opposite signs shown for example, in *Figure 3.7*, could be reduced if the dislocations came close together, because the overlapping of compressed and dilated zones would diminish the local strains. In fact the potential energy gradient between the dislocations manifests itself as a mutual attraction; however, if the dislocations on the slip plane were of equal signs they would repel, somewhat like electric line charges of equal signs.

The state of stress due to dislocations in homogeneous, elastically isotropic, bodies can be evaluated by classical methods of elasticity. The results may be applied, with certain modifications, to crystals. Apart from the elastic anisotropy of most crystals, which we shall disregard, the calculations cannot be applied to a small volume contained within a cylinder of a few atomic spacings in diameter

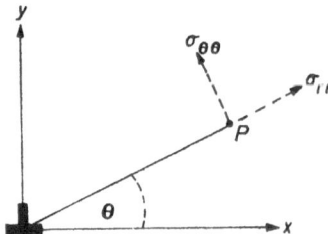

Figure 3.9. Stresses at a point in the crystal due to an edge dislocation lying along the z-axis

centred on the dislocation core, for in view of the large strains at the core, Hooke's Law, assumed in the calculations, is there no longer valid.

For an edge dislocation lying along the z-axis of a Cartesian co-ordinate system (*Figure 3.9*) the non-zero components of the stress tensor are

$$\left.\begin{aligned}
\sigma_{xx} &= -D.y \frac{(3x^2 + y^2)}{(x^2 + y^2)^2} & \sigma_{yy} &= D.y \frac{(x^2 - y^2)}{(x^2 + y^2)^2} \\[2mm]
\sigma_{zz} &= \nu(\sigma_{xx} + \sigma_{yy}) & \tau_{xy} &= D.x \frac{(x^2 - y^2)}{(x^2 + y^2)^2}
\end{aligned}\right\} \quad (3.8)$$

where $D = Gb/2\pi(1 - \nu)$, G being the shear modulus, b the Burgers vector, and ν Poisson's ratio. In cylindrical co-ordinates one obtains

$$\left.\begin{aligned}
\sigma_{rr} &= -(D/r) \sin \theta \\
\sigma_{\theta\theta} &= -(D/r) \sin \theta \\
\tau_{r\theta} &= (D/r) \cos \theta
\end{aligned}\right\} \quad (3.9)$$

54

with σ_{zz} equal to $\nu\,(\sigma_{rr} + \sigma_{\theta\theta})$. For a screw dislocation lying along the z-axis one obtains, as the only non-zero component of the stress,

$$\tau_{\theta z} = Gb/2\pi r \qquad (3.10)$$

which is independent of θ.

3.4 Tilt and Twist Boundaries in Crystals

As a consequence of their stress fields dislocations will exert forces on one another, and will tend to rearrange themselves in such a manner that the elastic energy of the crystal is reduced. Loss of dislocations by mutual annihilation may result, as well as the formation of relatively stable dislocation arrangements.

Referring to *Figure 3.9*, for example, edge dislocations of the same sign as the one at the origin will be attracted towards the y-axis if their x-co-ordinates are less than their y-co-ordinates; for τ_{yx} is then negative (equation 3.8). Hence, edge dislocations lying above the line $x = y$, which makes an angle of 45° with the x-axis, will tend to be swept into alignment along the y-axis. Once they have arrived in this position they in turn will ' rake in ' positive dislocations located close to them and above a similar line inclined by 45° to the axis, at the same time repelling nearby edge dislocations of negative signs. This discrimination and segregation process is therefore autocatalytic, and a dislocation wall or boundary may form. If the separation between dislocations in the wall is large compared with the lattice spacing, b, the parts of the crystal on opposite sides of the wall become tilted with respect to one another by an angle $\alpha = b/d$ (*Figure 3.10a*). In a bent crystal in which the process has been allowed to reach completion, so that only the excess dislocations of one sign necessary to maintain the bent shape remain, the dislocations will be arranged in tilt boundaries, approximately at right-angles to the slip planes, as indicated in *Figure 3.10b*.

If the radius of curvature R of the bent crystal is assumed to be large compared with the crystal thickness, and the dislocation walls are taken to be spaced a distance u apart

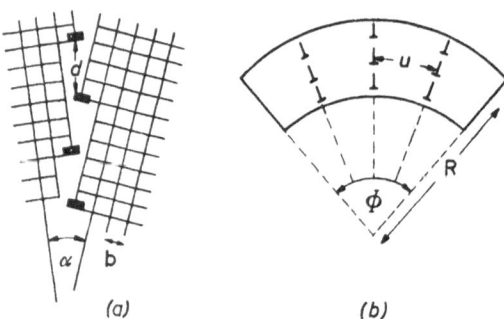

Figure 3.10.   Structure of: (a) a tilt boundary;
(b) a bent, polygonised, crystal

then, as each wall contributes a tilt b/d to the angle ϕ

$$(R\phi/u)b/d = \phi \tag{3.11}$$

The density of dislocations, given by the number intersecting unit area of crystal perpendicular to the glide plane, is

$$\rho = 1/ud$$

so that

$$\rho = 1/bR \tag{3.12}$$

This equation enables one therefore to estimate the density of excess dislocations of one sign introduced into a crystal in the process of bending. On account of the polygonal shape assumed by the crystal as a result of this process, it is often referred to as ' polygonisation '. *Plate I* shows the formation of ' cell ' walls or ' sub-boundaries ' by a process of polygonisation in a grain of deformed polycrystalline

56

*Plate I. Glide polygonisation in 99·999 per cent pure polycrystal-
line magnesium after deformation at 200° K*

magnesium. The glide planes intersect the surface in the slip lines running approximately diagonally from the right lower corner; the almost vertical sub-boundaries appear to have collected dislocations from the slip planes.

More complex mutual displacements of parts of crystals can be described by other arrangements of dislocations. Thus in *Figure 3.11* part of a crystal below the glide plane *ABCD* is held stationary, while the upper part is displaced by slip due to the introduction of a set of screw dislocations

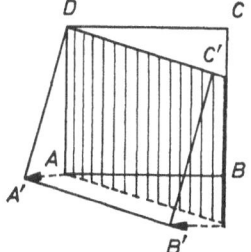

Figure 3.11. Formation of a twist boundary between parts ABCD and A'B'C'D of a crystal

parallel to *AD*. It becomes lozenge-shaped in the process. If now a second set of screw dislocations is introduced, this time parallel to *AB*, the upper part of the crystal regains its previous form. The deformation with the final position of the upper part at *A'B'C'D*, is equivalent to displacing the two parts of the crystal on opposite sides of the glide plane by twisting about an axis perpendicular to the slip plane. The resulting 'twist' boundary consists of a crossed grid of screw dislocations.

3.5 The Energy of Formation of Dislocations

The energy required to form an edge dislocation in a homogeneous, isotropic, elastic material may be obtained by evaluating the work necessary to displace the two surfaces formed by cutting a cylinder of radius r_1, and unit length, by an amount b, as indicated in *Figure 3.12*. We shall consider the energy stored in the entire volume of the cylinder except for a narrow core region of radius r_0 centred on the dislocation, where the displacement will be significantly less than b. This core will be discussed separately.

Now the shear stress on the slip plane at a point P a distance r from the origin will rise from zero at the time when the cylinder edges at A and A' coincide, to its full value $\tau = Gb/2\pi (1 - \nu)r$ (equation 3.9) when the dislocation is already situated at O. The mean stress in the course of introducing the dislocation will therefore be $\frac{1}{2}\tau$, and the

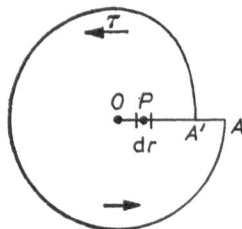

Figure 3.12. Formation of an edge dislocation in a cylinder of radius r_1 and unit length

work required to effect the displacement b over a strip of unit length and width $\mathrm{d}r$ is

$$\mathrm{d}W = \frac{1}{2} \left[Gb^2/2\pi (1 - \nu)\right] \mathrm{d}r/r$$

so that the total work to form a dislocation of unit length is

$$W = \left[Gb^2/4\pi (1 - \nu)\right] \ln (r_1/r_0) \qquad (3.13)$$

The energy of formation per segment of length b is

$$w = bW. \qquad (3.14)$$

If the core radius is taken to be about $4b$, and the range of the stress field of a dislocation is assumed to be rather less than the mean distance between dislocations in moderately deformed crystals, so that $r_1 \approx 1000\ b$, one obtains

$$W \approx \frac{1}{2}Gb^2 \qquad (3.15)$$

This estimate is not very sensitive to variations in the ratio r_1/r_0, which appears in equation 3.13 only as a logarithm.

An upper limit to the energy stored in the core can be obtained by considering it to be so heavily distorted that its structure resembles that of a liquid. The energy within it cannot then exceed the latent heat of melting within a tube of molten crystal having the same volume as the core. On evaluating this one finds this energy to be negligible compared with W, which may therefore be regarded as the total energy of formation per unit length of edge dislocation. By analogy with the two-dimensional ' surface tension ', W is sometimes referred to as the ' line tension ' or ' line energy ' of the dislocation. The above treatment may be applied to screw dislocations; similar results are obtained.

Now in copper, for example, w (equation 3.14) is found to be about $2\frac{1}{2}$ eV, and since a dislocation must be many lattice spacings long to form a loop or arc enclosing a slipped area, the energy of formation of such loop or arc would be so high that generation of dislocations by thermally activated processes is impossible. Exceptions may occur under rather special conditions, for example in atomically thin wedges of metal, bounding holes produced by electropolishing. In such a case the least length of dislocation required to span the wedge near its edge may be only a few interatomic spacings, and in view of the thinness of the wedge the elastic energy stored within it may be very considerably less than that given by equation 3.13.

Dislocation-free crystals of germanium, silicon and other elements have been made by special methods; in the case of metals they can at present be produced only in the form of extremely fine whiskers. In general, dislocations are invariably present in crystals, and they can act in a number of ways as sources of new dislocations. One of these mechanisms we shall now consider.

3.6 The Frank–Read Source

The force F acting on unit length of dislocation due to a shear stress τ in the slip plane can be found by considering the movement of an edge dislocation a distance dr in the direction of its Burgers vector **b**. The slip displacement over this area is b, and the force $\tau . dr$; one therefore obtains for the energy expended:

$$F . dr = (\tau . dr)b$$

or

$$F = \tau b \qquad (3.16)$$

The same result would be obtained for screw dislocations.

We are now in a position to examine the effect of a shear stress in the slip plane on the shape of a dislocation. In *Figure 3.13*, let AB represent a short segment of length x of a dislocation firmly pinned at its two ends, e.g. at walls

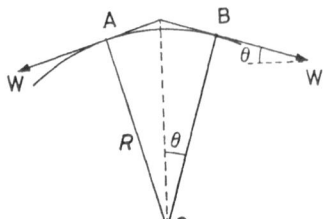

Figure 3.13. Equilibrium of a dislocation pinned at its ends, under a shear stress τ

of the crystal. Because of its line tension the dislocation will bow out in the slip plane under the applied shear stress until it attains an equilibrium position. This occurs when the force $(\tau b) x$ due to the stress is balanced by the opposing component of the line tension $2W\sin \theta$.

With θ taken small, also noting that $x = 2R\theta$, the condition of equilibrium then yields

$$\tau b = W/R \qquad (3.17)$$

and substituting for W from equation 3.15, the radius of curvature is found to be inversely proportional to the stress, and given by

$$R = \tfrac{1}{2}Gb/\tau \qquad (3.18)$$

Thus in the absence of a stress the dislocation will be

stretched straight. The radius of curvature will diminish as the stress is increased; consecutive stages are indicated by 0, 1 and 2 in *Figure 3.14.*

When the dislocation has become semicircular and the stress is increased further it is geometrically impossible to satisfy equation 3.18, and the dislocation becomes unstable, first developing re-entrant lobes, then breaking up into two parts due to annihilation of parts of the coalescing lobes of opposite sign which, assuming the dislocation was initially of the edge type, are screw segments, as shown in the

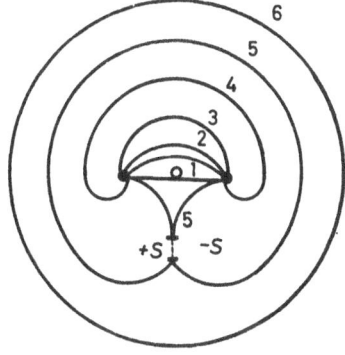

Figure 3.14. Consecutive stages in the expansion of a dislocation pinned at two points, under the action of an increasing shear stress

diagram. The pretzel-shaped part, numbered 5, straightens to become a dislocation loop, while the remaining, cusp-shaped, dislocation reverts to 1, thus recreating the source. The mechanism, first suggested by F. C. Frank and W. T. Read, is therefore capable of operating again, throwing off more loops. Various obstacles, such as 'forest' dislocations intersecting the slip plane containing the source, may interfere with it and thus arrest the 'mill' after a few loops have been generated. Evidence of the existence of Frank–Read sources in crystals has been obtained, though other mechanisms leading to an increase in the dislocation density can be imagined, and are probably of more frequent occurrence. The pinning points may again be dislocations on slip planes intersecting the glide plane of the source dislocation, though impurities and other lattice defects could also provide effective local locks.

A crystal containing dislocations pinned at points spaced approximately l apart would therefore show a fairly abrupt onset of plasticity when R (equation 3.18) becomes equal to $\frac{1}{2}l$, which is the criterion for activating the sources. The flow stress of the crystal would therefore be given by

$$\tau = Gb/l \qquad (3.19)$$

In well-annealed metal crystals the free length of dislocations may be several microns (*Plate II*). If, for example, one takes $l = 5$ μm, equation 3.19 yields for copper a flow stress of about 25 kg/cm², which is close to the experimentally determined value. On comparing equations 3.5 and 3.19 one observes that the flow stress of a real crystal is by a factor of approximately $10b/l$ smaller than that of dislocation-free ideal crystals.

3.7 Preferred Slip Systems and Miller Indices

Some reference to slip systems has already been made in section 2.2, without however a consideration of the crystallographic nomenclature used to describe lines and planes in crystals. We shall now discuss the specification of lines and planes in crystallographic terminology.

Now, it is obviously desirable to use a reference system of axes bearing some relation to the crystal morphology. *Figure 3.15* shows such a set for a triclinic crystal where the lengths of edges of the unit cell a, b and c and the angles between the faces are all unequal. The units in which the axes are measured are multiples of the lattice lengths. The shaded plane, for example, is specified by (111), and the same set of indices would be used to describe the shaded plane in the face-centred cubic crystal shown in *Figure 2.1*.

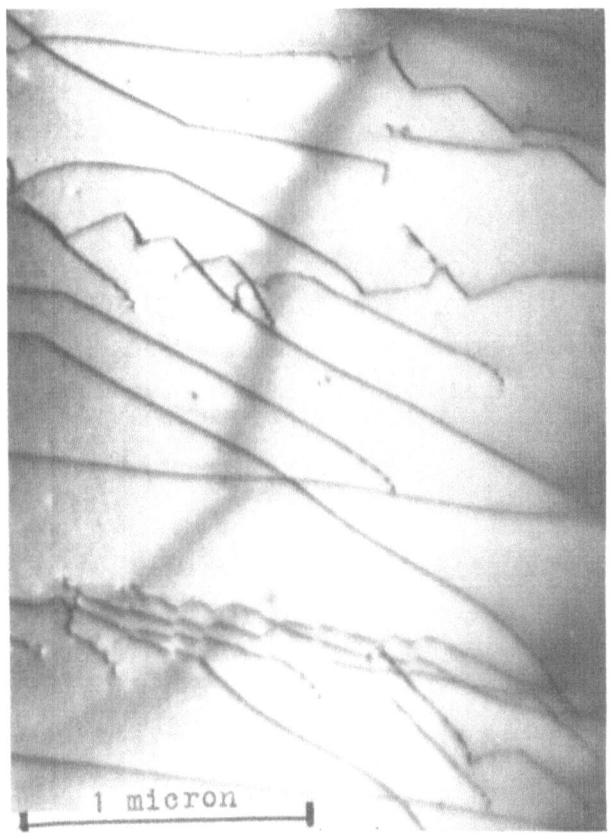

1 micron

Plate II. Electron transmission micrograph of dislocations in a slightly deformed copper foil less than ½ μm thick. The slip plane containing the dislocations is inclined at a few degrees to the plane of the foil, and some dislocations are seen to terminate at points where they meet one of the foil surfaces

The set of numbers in parentheses are known as the Miller indices of the plane, and for any given plane they are obtained as follows. Consider a plane which, to give a specific example, makes intercepts $1a$, $0 \cdot 5b$ and $3c$ on the co-ordinate axes. We now take the set of inverses of these numbers, i.e. $1/1$, $1/0 \cdot 5$ and $1/3$ and multiply all of them by the smallest number which will convert all the fractions to units. In the present case multiplication by 3 yields $(3, 6, 1)$, which are the required Miller indices. Similarly, the shaded plane in the sodium chloride crystal (*Figure 2.1b*), which

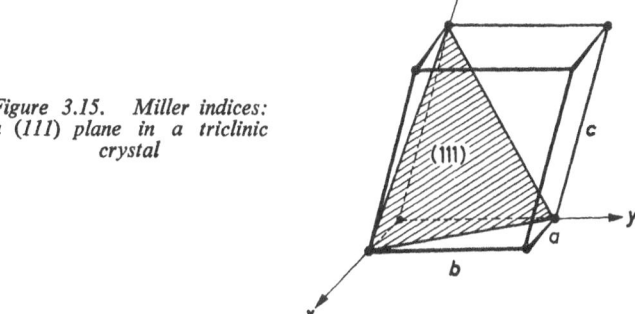

Figure 3.15.   Miller indices: a (111) plane in a triclinic crystal

intersects the x, y, z axes at 1, ∞, 1, has Miller indices $(1, 0, 1)$. If the intercepts on the axes are negative a minus sign is placed above the number, e.g. $(2, \bar{3}, 0)$.

In order to specify the direction of a line it is necessary to give the co-ordinates of two points. However the convention is to assume that the line passes through the origin, so that only one point has to be given explicitly. Again, using a definite example, a line passing through the point with co-ordinates $1a$, $0 \cdot 5b$ and $3c$ would be described by $[2, 1, 6]$, the numbers being proportional to the set 1, $0 \cdot 5$, 3, but without fraction or common factor. As can be seen,

63

a square bracket is used. Minus signs are again placed above the numbers; the convention is the same as with the planes.

The Miller index notation is particularly useful in the case of cubic crystals; the units of length, a, b and c are then equal. The Miller index of a plane (h, k, l) is then the same as that of the normal to the plane, $[h, k, l.]$ The angle θ between two planes is equal to that between their respective normals, i.e. between the lines $[h_1, k_1, l_1]$ and $[h_2, k_2, l_2]$, and is given by the equation

$$\cos \theta = \frac{h_1 h_2 + k_1 k_2 + l_1 l_2}{(h_1^2 + k_1^2 + l_1^2)^{\frac{1}{2}}(h_2^2 + k_2^2 + l_2^2)^{\frac{1}{2}}} \qquad (3.20)$$

It is readily checked, for example, that the planes (101) and ($\bar{1}$01) are mutually perpendicular, for the numerator in equation 3.20 is zero. For the angle between [001] and [101] one has $\cos \theta = 1/2^{\frac{1}{2}}$, and hence $\theta = 45°$.

Equation 3.20 is also useful if it is desired to establish whether a given line, for example [11$\bar{2}$], is parallel to a certain plane, e.g. (111). The scalar product [11$\bar{2}$].[111] given by the numerator in equation 3.20 must then be zero, as is in fact found to be the case.

The preferred slip systems in the face-centred cubic metals, which include Al, Ag, Au, Cu, Ni, Pb, Pd, Pt, Rh and others, are {111}, <110>, which are also operative in the cubic structure of the tetravalent homopolar crystals of diamond, germanium and silicon. In alkaline halides the slip planes and directions are {110}, <110> as indicated in *Figure 2.1*. The slip planes of body-centred cubic metals such as iron and tungsten are not unique, {112} and {110} may be operative; the slip direction remains however <111>. The braces and angular brackets indicate planes and lines of certain types respectively. Thus {112} is the shorthand notation for (112), (211), (121), (1$\bar{1}$2), etc.

4

DISLOCATION KINETICS AND LATTICE DEFECTS

4.1 Close Packing and Partial Dislocations

Structures typical of most metals can be regarded as consisting of closely packed spherical ions in contact. Only two crystal structures can be obtained in this manner, as can be seen by considering possible ways of stacking close-packed layers of the type shown in *Figure 4.1*. The numbered arrows indicate possible positions at which ions of adjacent sheets could be centred. If, for example the stacking sequence is 1, 2, 3, 1, 2, 3, 1 . . . the resulting lattice is face-centred cubic, as is apparent from *Figure 4.1b*, while if the stacking is such that only two positions are utilised, i.e. 1, 2, 1, 2, 1 . . . the close-packed hexagonal structure is obtained. Hexagonal metals are less common than face-centred cubic ones; they include among others beryllium, magnesium, zinc and cadmium.

In the course of plastic deformation the close-packed planes slip over one another without disturbing the stacking sequence. Thus an ion located at a certain ' 2 ' position in *Figure 4.1a* would be transferred to a crystallographically equivalent adjacent ' 2 ' position by the passage of a dislocation. Possible Burgers vectors must therefore join equivalent sites. However, a sheet of ' 2 ' ions could be imagined to slip over a ' 1 ' sheet by a sequence in which the transfer of ions from positions A to B (*Figure 4.1a*) occurred via ' 3 ' positions, as indicated by the Burgers vectors $\mathbf{b_1}$ and $\mathbf{b_2}$ into which $\mathbf{b}$ may be considered to have been resolved. While the vector sum of $\mathbf{b_1}$ and $\mathbf{b_2}$ is equal to $\mathbf{b}$, it is clear that

$$\mathbf{b_1^2} + \mathbf{b_2^2} < \mathbf{b^2} \qquad (4.1)$$

since the angle between $\mathbf{b}_1$ and $\mathbf{b}_2$ is 120°. Assuming elastic isotropy, it follows from equation 3.15 that replacement of the dislocation with Burgers vector $\mathbf{b}$ by two 'partial'

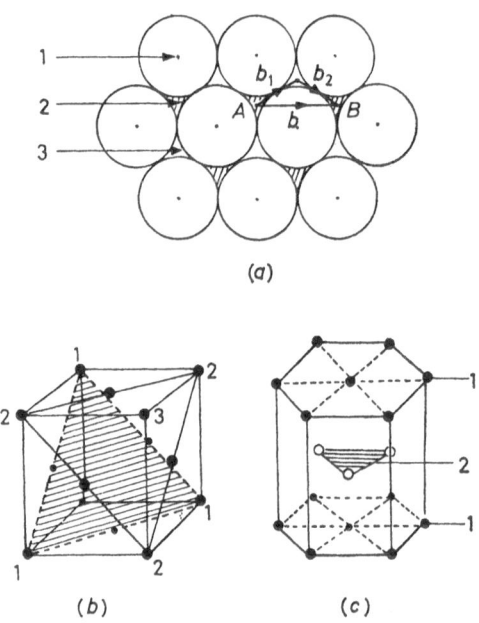

(a)

(b)   (c)

Figure 4.1. Close-packed structures. In (a) centres of a further layer of close-packed spheres could be placed either into positions 2 or into positions 3. The Burgers vector $\mathbf{b}$ can be split vectorially into $\mathbf{b}_1 + \mathbf{b}_2$. The face-centred cubic and hexagonal close-packed structures are obtained by stacking close-packed sheets in the sequences 1, 2, 3, 1, 2, 3, 1 . . . and 1, 2, 1, 2, 1 . . . respectively

dislocations with Burgers vectors $\mathbf{b}_1$ and $\mathbf{b}_2$ respectively would result in a reduction of the elastic energy of the crystal. The splitting of a glide dislocation into two partials is therefore energetically favoured here; that it does in fact occur is experimentally well documented. As energy would be required to recombine two partials, a force of repulsion must exist between them. However, they can move apart only over a definite, fixed distance, for in the process of separation a stacking fault is formed between them, and the available energy becomes expended. The stacking fault tends to draw the dislocations together, somewhat like a soap film possessing surface energy; the balance of attractive and repulsive forces acting on the partials will determine the equilibrium separation between

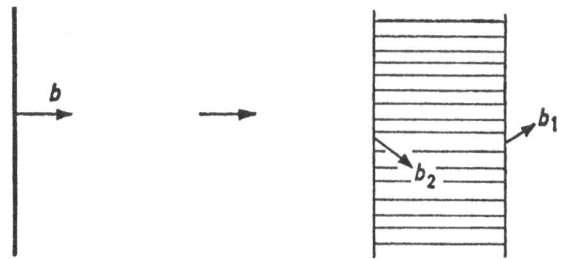

*Figure 4.2. Splitting of an edge dislocation of Burgers vector **b** into two parallel partial dislocations of mixed character and Burgers vectors **b₁** and **b₂**. The shaded region represents the stacking fault*

them. Typical values of the width of the stacking-fault ribbon are about $1b$ for aluminium and $10b$ for copper, but larger values may occur in certain alloys such as stainless steel, in graphite and other crystals. That a stacking fault must form between the two partials (*Figure 4.2*) may be seen by considering that the passage of the first partial dislocation over a certain area of the slip plane of a face-centred cubic metal, for example, will change the stacking

sequence of 1, 2, 3, 1, 2, 3 to 1, 2 | 3, 1 ↓ 3, 1 | 2, 3, which

contains a 3, 1, 3, 1 sandwich characteristic of hexagonal crystals. The second partial dislocation completes the slip

67

process, thereby restoring correct stacking. Similarly, stacking faults having face-centred cubic structure can be introduced into hexagonal metals in this manner.

Splitting of dislocations into partials is of considerable significance in relation to certain features of the plastic behaviour of crystals. We see for example that screw dislocations which, as we know, have a radially symmetric stress field and do not 'carry' an excess plane of atoms, could cross-slip from one glide plane to an intersecting one, as indicated in *Figure 4.3*. In aluminium and in body-centred metals where glide dislocations are little or not at

Figure 4.3. Cross-slip of a screw dislocation

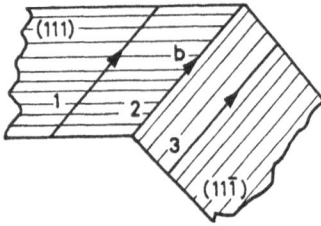

all dissociated this process appears to take place relatively readily. Dissociated screw dislocations cannot cross-slip until they have recombined, at least over a certain length; hence rather special circumstances must prevail to facilitate this. Cross-slip is then more difficult and relatively rare.

4.2 Jogs and Point Defects

As can be seen from *Figure 4.4* slip on intersecting glide planes leads to the formation of steps on the dislocations, known as 'jogs'. The sharp bend in the dislocation at the jog distorts the lattice. A jog will therefore exercise a frictional drag on the dislocation containing it.

Now, remembering that the Burgers vector of a disloca-tion is the same at all of its parts, hence also at the jog, the latter can be seen (*Figure 4.5*) always to have edge character. In the case of the edge dislocation, denoted by E, the glide

68

plane of the jog contains its Burgers vector, so that the jog can move with the dislocation in the direction *AB* without great difficulty. As will be seen, however, it cannot move at right-angles to the direction of *AB* without ploughing up the lattice and forming vacant lattice sites or interstitial ions in the process.

Figure 4.4. Formation of jogs on dislocations through intersection by dislocations moving on another slip system

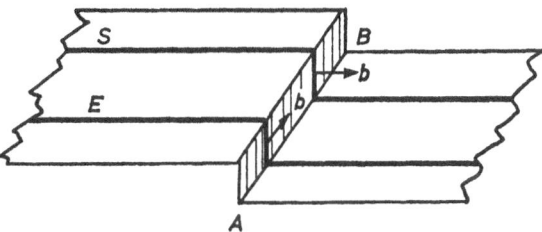

Figure 4.5. Jogs in edge and screw dislocations having Burgers vectors along and at right-angles to the direction of motion AB respectively. The jog in the screw dislocation has edge character and can move with the dislocation along AB only non-conservatively

By contrast, the Burgers vector of the jog in the screw dislocation *S* is perpendicular to the direction of movement *AB*. The jog could therefore migrate along the dislocation but not with the dislocation line, along *AB*, for in that case it would be forced to migrate at right-angles to its Burgers vector and, again, it would generate vacancies or interstitials. The mode of migration of a jog or dislocation along its slip plane is termed 'conservative'; movement resulting in the formation of point defects is 'non-conservative'.

How point defects form through jog movement can be seen by considering the non-conservative (a) upward and (b) downward displacement of an edge dislocation in a monatomic sheet of crystal, illustrated in *Figure 4.6*; the slip plane is horizontal in both cases. In (a) the ion located at the lattice point '1' jumps away into an interstitial position '2', and the dislocation consequently climbs to

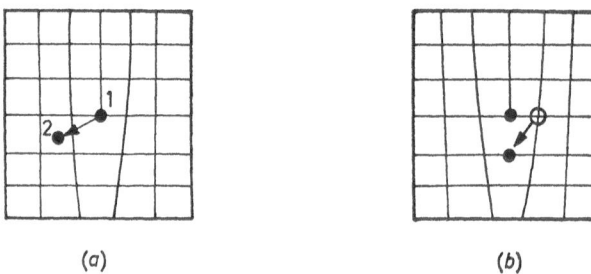

(a)   (b)

Figure 4.6. Formation of (a) an interstitial and (b) a vacancy by non-conservative ' climb ' of an edge dislocation

the slip plane above the original one; in (b) an ion jumps below the centre of the dislocation, extending the latter and leaving a vacancy adjacent to it. The force on the jog arises from the stress acting on the dislocation containing it, so that the stress in the slip plane of the glide dislocation provides the energy required to form chains of point defects as the jog is being dragged along non-conservatively. Thermal activation may assist this process at elevated temperatures. Dissociated jogs may behave in a more complex manner than undissociated ones; there is some evidence which suggests that conservative movement may be difficult or even impossible in such cases.

70

4.3 Obstacles to Glide

A dislocation moving on its slip plane will be intercepted by obstacles, primarily by the ' forest ' dislocations threading its glide plane. In order to be able to move over significant distances the dislocation must either cut through the forest dislocations, with the resulting formation of jogs in the forest and glide dislocations, or the stress must ' extrude ' it past the forest dislocations as shown in *Figure 4.7*. In the latter case, when the arcs between the obstacles *A* and *B* have become almost semicircular the segments of opposite signs will pinch off, in a manner similar to that described in

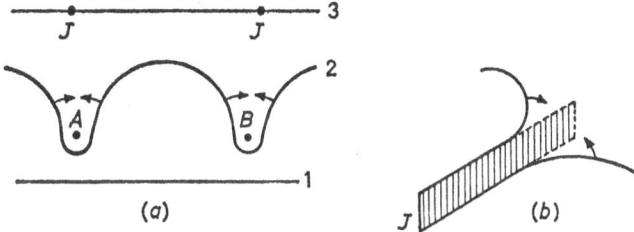

Figure 4.7. Obstacles to dislocation movement : (a) forest dislocations A and B ; (b) a large jog, giving rise to a dipole

the discussion of the Frank–Read source. The dislocation, which may become jogged in the process, will then be able to move along under the applied shear stress. The jogged loops remaining threaded around the forest dislocations would contract under their own line tension and disappear, in certain cases leaving stable jogs in the forest dislocations. If a dislocation contains a ' large ' jog, i.e. a dislocation segment many interplanar spacings long (*Figure 4.7b*), an elongated dipole may be left in its wake.

Whether in the course of plastic deformation the forest is intersected as indicated in *Figure 4.7a*, or whether a more direct cutting of dislocations occurs may depend on the material and specific conditions, for example certain stress and temperature combinations. In the direct cutting process two jogs would form simultaneously with an expenditure of energy equal to an appreciable fraction of Gb^3 (equation 3.14); thermal activation could assist the process and a high temperature sensitivity of the flow stress would be expected. In the extrusion mechanism the work effecting ' extrusion ' is done by the applied stress, and no significant thermal activation would occur. The relatively low temperature sensitivity of the flow stress of annealed metal crystals suggests that an extrusion process occurs as a rule.

4.4 The Velocity of Jogged Dislocations

A dislocation moving in a crystal otherwise free from imperfections should be able to accelerate to a velocity equal to an appreciable fraction of that of sound in the material. In general, however, they move comparatively slowly, as has been found by direct observation by electron transmission micrography and by etch-pit methods. The relatively low mean velocities must be due to the periodic arrest of the dislocation at obstacles, to the drag of jogs and that of impurities which, by interacting with the stress field of the dislocation, tend to pin it, or to combinations of all three effects.

We shall consider the drag of jogs, for these are invariably formed if the material is plastically strained, irrespective of chemical purity. Vacancy generating jogs on screw dislocation will be taken as a specific illustration.

Now *Figure 4.8* shows a jogged screw dislocation with jogs spaced l_1, l_2, etc. apart, bowed out into arcs of radii of curvature R by an effective shear stress τ. The force on the jog J arises from the line tension W of the dislocation, and its component F perpendicular to the Burgers vector and along the direction of motion is

$$F = W(\sin\theta_1 + \sin\theta_2) \tag{4.2}$$

while the component perpendicular to F, tending to move the jog conservatively along the dislocation is

$$F' = W(\cos\theta_1 - \cos\theta_2) \qquad (4.3)$$

In general therefore the jog will move sideways as well as forward, the respective velocity components depending on F and F', and on the magnitudes of the energies required to effect the displacements. In the direction of F the motion is non-conservative; a point defect has to be formed for each interatomic displacement in this direction. Along the dislocation the jog can move conservatively, i.e.

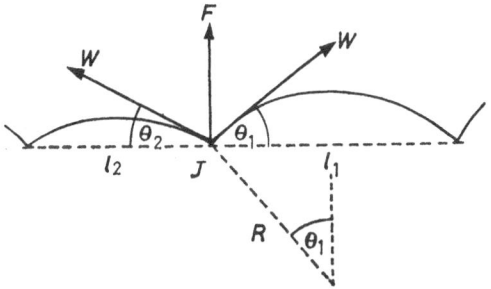

Figure 4.8. A screw dislocation dragging a jog J

by increasing l_1 at the expense of l_2, without migrating in the direction of F. Such a conservative displacement could not continue if the jog ran into parts of the dislocation not of purely screw character.

Although a jog may eventually run off the dislocation, out of the crystal, or become annihilated by encountering a jog of opposite sign, dislocations will acquire new ones in the course of their passage through the forest. At any given instant a dislocation will therefore be subjected, in general, to a drag arising from non-conservative jog displacements. We shall therefore examine the drag of an array of 'non-conservative' jogs on a screw dislocation, assuming the jogs to be a fixed distance apart; this static distribution will be taken to be equivalent to the actual, dynamic one, where jogs are both formed and lost. As these ' equivalent' jogs move with the dislocation in the direction F their velocity and that of the dislocation must be the same. Now, if the energy of formation of a point defect is Q_0 then, since the work done by the line tension of the dislocation in the process of forming the point defect is Fb, where we take b to be the lattice spacing in the direction of F, the frequency v_1 of forward jumps will be given by the Boltzmann relation

$$v_1 = v_0 \exp[- (Q_0 - Fb)/kT] \qquad (4.4)$$

for a thermally activated rate process in which the energy barrier Q_0 is reduced by the work contributed to point defect formation by the stress. Here v_0 is an atomic vibration frequency of the order of 10^{12} per second, $\mathbf{k}$ is Boltzmann's constant and T the temperature in °K. Jumps in the reverse direction will be impeded by the stress, and the frequency for backward jumps will be

$$v_2 = v_0 \exp[- (Q_0 + Fb)/kT] \qquad (4.5)$$

The net forward jump rate will therefore be

$$v = v_0 \left[\exp\left(-\frac{Q_0 - Fb}{kT}\right) - \exp\left(-\frac{Q_0 + Fb}{kT}\right) \right] \quad (4.6)$$

The second term is generally negligible compared with the first one; if, however, both are taken into account we can write equation 4.6 in the form

$$v = 2v_0 \exp(- Q_0/kT) \sinh(Fb/kT) \qquad (4.7)$$

Since the jog, and hence the dislocation, moves a distance b in the direction F in each successful jump, the dislocation velocity is

$$u = vb = 2v_0b \exp(- Q_0/kT) \sinh(Fb/kT) \qquad (4.8)$$

and this is linear in F if $Fb \ll kT$.

74

The shear stress τ can be introduced into this equation as follows. On writing $l_1 = l_2 = \ldots = l_j$, where l_j is the equivalent or 'effective' jog spacing, one obtains, noting that then $\theta_1 = \theta_2 = \theta$ (*Figure 4.8*)

$$2W\sin\theta = F; \quad R\sin\theta = \tfrac{1}{2}l_j \qquad (4.9)$$

which yields with equations 3.15 and 3.18:

$$F = \tau b l_j \qquad (4.10)$$

so that

$$Fb/\mathbf{k}T = \tau b^2 l_j/\mathbf{k}T \qquad (4.11)$$

The product $b^2 l_j$ is known as the 'activation volume'.

5

WORK-HARDENING, RECOVERY AND CREEP

5.1 The Stress-Strain Curve of Single Crystals

We consider a cube-shaped single crystal of length of side L_0 containing one dislocation of Burgers vector **b**, as in *Figure 3.5*. When the dislocation has passed through the entire crystal the shear strain will be b/L_0. However, if it moves into the crystal from the left over a distance $x < L_0$, the shear strain will amount to only $(b/L_0)(x/L_0)$. When n dislocations have moved along parallel glide planes into the crystal the shear strain would be $\frac{1}{2}nb/L_0$, taking the mean value of x/L_0 for a uniform distribution of dislocations to be $\frac{1}{2}$. The ratio n/L_0^2 is the density ρ of glide dislocations, so that the shear strain due to them is

$$\gamma = \tfrac{1}{2}\,\rho\,bL_0 \tag{5.1}$$

Provided that the number ρ' of forest dislocations threading unit area of the active glide planes under consideration remains constant, the mean distance between points at which forest dislocations intersect an active slip plane will be fixed and equal to $(1/\rho')^{\frac{1}{2}}$. According to equation 3.19 the flow stress will therefore remain constant, independent of the strain, the distance between pinning points on the glide dislocations being determined by ρ'.

Constancy of the density of forest dislocations can be maintained in the early stages of deformation of single crystals of many metals which have been well annealed. The deformation then proceeds without significant work-hardening up to strains depending on the perfection and dimensions of the crystal and its orientation with respect to the tensile axis, factors which may influence the onset of glide on intersecting slip systems. This 'easy glide', generally referred to as 'stage 1' is then followed by 'stage 2' of intense work-hardening (*Figure 5.1*); here interaction of dislocations on intersecting slip systems leads to the formation of stable dislocation networks.

As the stress is increased new dislocations move into the crystal, preferentially in regions where they are least impeded by the stress fields of existing ones. 'Soft' spots in the

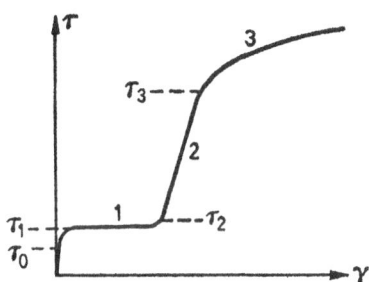

Figure 5.1. The work-hardening curve characteristic of single crystals. All three stages may not always be realisable

crystal, containing a 'below average' density of dislocations therefore become less soft as the deformation proceeds, and the network tends to assume similar mesh dimensions and geometrical features throughout the crystal. The mesh diminishes with increasing stress, without however significant changes in its shape. Groups of large loops as shown in *Figure 5.2* may therefore become superseded by sets of smaller ones as a result of the overlapping and interaction of dislocations moving from the periphery of the large loop into its interior.

Now if the slip distance at any stage of this parcellation process is L, the corresponding mean spacing between

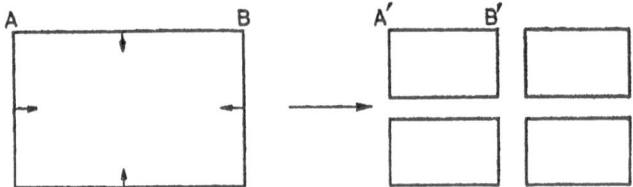

*Figure 5.2. Parcellation of a large loop into a set of four geo-
metrically similar ones. (Schematic)*

adjacent dislocations is d, and the separation of active slip
planes D, then the maintenance of geometrical similarity
suggests the relation

$$L = c_1 D = c_2 d \qquad (5.2)$$

or

$$L = (c_1 c_2 D d)^{\frac{1}{2}} = (c_1 c_2)^{\frac{1}{2}}/\rho^{\frac{1}{2}} \qquad (5.3)$$

where ρ is the dislocation density and c_1 and c_2 are constants
determining the geometry of the dislocation distribution.

*Figure 5.3. A disloca-
tion source S of edge
type in the field of
an edge dislocation
of the same sign,
also subjected to an
applied shear stress τ*

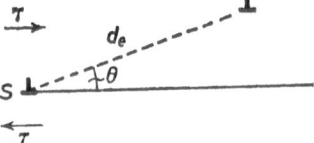

The origin of the constancy of the ratios of the dimensions L, D, d, expressed by equation 5.2, may be illustrated
as follows. Consider a dislocation source, e.g. of edge type,
l_s long, subjected to an applied shear stress τ in its slip
plane. Let it also be subjected to an opposing shear stress
due to an edge dislocation of equal sign fixed in the lattice
at a distance d_e from the source, as shown in *Figure 5.3*;
the angle θ is assumed to be small to avoid mathematical

difficulties. The source will operate if the shear stress complies with the equation

$$\tau - \frac{Gb}{2\pi \, (1 \, - \, \nu)d_e} \approx \frac{Gb}{l_s} \tag{5.4}$$

which follows from equation 3.19, assuming $d_e \gg l_s$. If τ is the flow stress at which a large number of sources become activated simultaneously then one can infer from equation 5.4 that

$$\tau \approx \frac{Gb}{l_s} \left[1 \, + \, \frac{l_s}{2\pi \, (1 \, - \, \nu) \, d_e} \right] \tag{5.5}$$

and that the flow stress over a small range of source lengths would not vary significantly provided l_s/d_e remained approximately invariant. If one takes for the distribution of source lengths and corresponding spacings d_e

$$l_s = l_0 \, (1 \, + \, x), \qquad d_e = d_0 \, (1 \, - \, ax), \qquad | \, x \, | \, \ll 1$$

with

$$a = 2\pi \, (1 \, - \, \nu) \, d_0/l_0$$

then τ is independent of x, and the ratio l_s/d_e is very nearly constant. Conservation of geometrical network similarity is probably a consequence of the tendency of dislocation movement and formation to occur so as to introduce the least elastic strain energy into the crystal for any given plastic strain increment.

Now returning to *Figure 5.2*, we note that if the slip distance L is approximately equal to a characteristic dimension of the rectangles, such as AB and $A'B'$, then by equation 5.1 the shear strain would be

$$\gamma \approx \tfrac{1}{2} \, (1/Dd) \, bL \tag{5.6}$$

If the distance between forest dislocations, and hence between ' hard ' immobile pinning points on any given dislocation is taken to be $(Dd)^{\frac{1}{2}}$ (equation 5.3), then one obtains from equation 3.19, irrespective of whether it is applied to a Frank—Read source or to the 'extrusion' mechanism considered in section 4.3, the relation

$$\tau = Gb\rho^{\frac{1}{2}} \tag{5.7}$$

The product Dd has here been equated to $1/\rho$. Equations 5.3, 5.6 and 5.7 yield the linear work-hardening law

$$\tau/\gamma = h; \qquad h = G/\frac{1}{2}(c_1 \, c_2)^{\frac{1}{2}} \tag{5.8}$$

characteristic of ' stage 2 '.

The values of the constants of proportionality c_1 and c_2 cannot be estimated without further detailed analysis. However, with reasonable values, for example $L = 1$ mm, $D = 10$ μm and $d = 1$ μm, corresponding to a dislocation density of 10^7 cm^{-2}, one obtains $c_1 = 100$ and $c_2 = 1000$, so that

$$h \approx G/150 \tag{5.9}$$

which is in good agreement with measured values, available mainly for close-packed metal crystals. The shear modulus varies only slowly with temperature, and h would therefore be expected to be rather temperature-insensitive; this again is confirmed by experiment.

When a certain dislocation density has been attained, characterised by the temperature-dependent shear stress τ_3 (*Figure 5.1*), the dislocations begin to interact sufficiently strongly to be able to rearrange themselves by glide poly-gonisation and cross-slip, thereby reducing the strain energy stored in the crystal. Mutual annihilation may also occur. These processes will take place already in the course of deformation above τ_3, and the rate of accumulation of dislocations will therefore be less than if such ' dynamic recovery ' were absent. The strain will now be due not only to slip by the dislocations present in the crystal; it will include contributions from dislocations which are no longer in the material. One may therefore write instead of equation 5.8

$$\gamma = \frac{\tau}{h}[1 + a\,(\gamma - \gamma_3)] \qquad (5.10)$$

where a is a constant which will in general increase with increasing temperature and decrease with increasing strain rate. The term $\gamma - \gamma_3$ is taken to be zero at strains below γ_3; it represents a contribution to the total strain by recovery processes. If one rewrites equation 5.10 in the form

$$\gamma = \tau/h' \qquad (5.11)$$

with

$$h' = h/[1 + a\,(\gamma - \gamma_3)] \qquad (5.12)$$

the coefficient of work-hardening h' in ' stage 3 ' of the work-hardening curve is seen to decrease uniformly to zero with increasing strain. The observed rapid decrease of the slope of the work-hardening curve as deformation proceeds in stage 3 is qualitatively in agreement with equation 5.12.

5.2 Work-hardening of Polycrystals

Easy glide is not observed in polycrystals, since operation of one glide system only would not allow all the grains to deform simultaneously and uniformly. This is apparent, for example from equation 2.4 which shows that the tensile flow stress

$$\sigma = \tau_{cr}/\cos\,\theta\,\cos\,\phi$$

would have to be exceedingly high for certain crystals, or grains, which would therefore fracture rather than deform, unless alternate more favourable slip systems were available. Hexagonal metals for example, in which the close-packed basal plane is the preferred glide plane, with other systems less easily induced to propagate slip, are appreciably less ductile than face-centred cubic crystals which have several non-parallel slip systems of the same type (section 3.7) to accommodate glide.

Although more than one slip system must operate in polycrystalline hexagonal metals to allow them to attain the observed, not insignificant, plastic strains without fracture, the basal planes are generally most active, with slip on other systems occurring mainly near grain boundaries. Stage 2 of rapid work-hardening, which involves dislocation interaction on intersecting systems, does not therefore seem to take place to an appreciable extent in hexagonal metal polycrystals, although it is observed in single crystals.

'Linear' hardening, corresponding to stage 2 in single crystals, is readily demonstrated in face-centred cubic metals provided they are sufficiently well annealed to have a tensile flow stress less than σ_3, which corresponds to τ_3 in single crystals. The magnitude of the coefficient of work-hardening $d\sigma/d\epsilon$ can be deduced from h (equation 5.9) as follows.

Consider a polycrystalline tensile specimen consisting of grains with a mean flow stress in shear τ. On deforming the specimen plastically by a tensile strain increment $d\epsilon$ the amount of work done per unit volume is $\sigma.d\epsilon$, and this must be equal to $\tau.d\gamma$, where $d\gamma$ is the mean 'equivalent' shear strain increment in the crystals. On equating these energies:

$$\sigma.d\epsilon = \tau.d\gamma$$

one obtains

$$\sigma/\tau = d\gamma/d\epsilon \tag{5.13}$$

Now the mean shear stress would be expected to be somewhat less than the maximum shear stress $\frac{1}{2}\sigma$; in fact it may be shown that

$$\tau \approx \tfrac{1}{3}\sigma \tag{5.14}$$

which, substituted into equation 5.13 yields

$$\gamma \approx 3\epsilon \tag{5.15}$$

Thus from equations 5.8, 5.14 and 5.15 one obtains,

$$\mathrm{d}\sigma/\mathrm{d}\epsilon = H \qquad (5.16)$$

where

$$H \approx 9h \qquad (5.17)$$

and the same equations could be used to adapt equation 5.10 for polycrystals. In view of the intracrystalline processes visualised in the derivation of equation 5.10 the modification derivable from it for the stress–strain relation of polycrystals would not be expected to have grain size as a variable. In fact, in face-centred cubic metals the stress–strain relation is relatively insensitive to grain dimensions. In hexagonal metals, however, the slip distance L_0 introduced into equation 5.1 should be equal to a substantial fraction of the grain diameter, for slip on non-basal planes would be relatively ineffective in preventing the passage of dislocations across an entire grain. We therefore use equation 5.1, appropriate to such a model of glide in the grains, together with equation 5.7, and obtain the ' parabolic ' work-hardening law

$$\tau^2/\gamma = 2G^2b/L_0 (1 + a) \qquad (5.18)$$

where the factor $(1 + a)$ has again been introduced to allow for the contribution to the strain by dislocations effectively removed by recovery processes. It should be noted that the dislocation density in equations 5.1 and 5.7 is the average density in the metal; individual grains will have different densities $\rho(L)$, depending in general on the grain dimension. If the shear strain is not to fluctuate unduly from grain to grain, so that coherency is maintained, equation 5.1 suggests that the product $\rho(L)$. L would be roughly constant. Thus the dislocation density would then tend to be below average in large grains and above average in small grains. There is experimental evidence in support of this conclusion.

On using equations 5.14 and 5.15 in equation 5.18, one obtains

$$\sigma^2/\epsilon = \chi \qquad (5.19)$$

with the temperature-dependent coefficient of 'parabolic' work-hardening equal to

$$\chi(T) = 54G^2b/L_0(1 + a) \qquad (5.20)$$

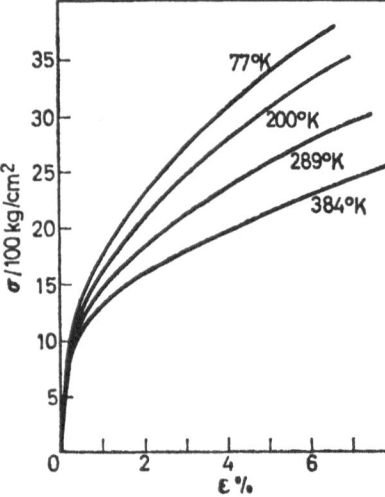

Figure 5.4. Tensile stress–strain curve of polycrystalline cobalt deformed at a strain rate of about 0.1 per cent per second

One may infer from equations 5.19 and 5.20 that if a series of well annealed polycrystalline specimens of the same material but different grain dimensions L_0 are subjected to tensile tests, then the flow stresses measured at the same strain will be proportional to $\chi^{\frac{1}{2}}$, and hence to $L_0^{-\frac{1}{2}}$. Such a proportionality, sometimes referred to as Petch's relation, has been studied extensively in steels. It explains the well-known opinion that fine-grained polycrystals are generally stronger than coarse-grained ones.

Figure 5.4 shows the tensile stress-strain curves of annealed polycrystalline cobalt at various temperatures, well within the range where the metal is close-packed hexagonal. The applicability of equation 5.19 is apparent from *Figure 5.5* in which the 'plastic' part of the curves,

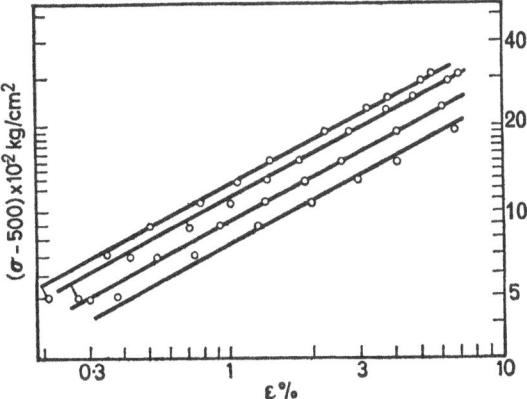

Figure 5.5. Log–log representation of the work-hardening curves shown in Figure 5.4. The slopes of all lines are ½

Figure 5.6. Temperature dependence of the coefficient of 'parabolic' work-hardening in polycrystalline cobalt. Measured points were obtained from Figure 5.5

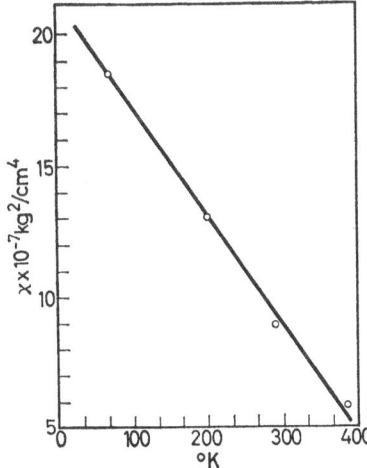

i.e. above $\sigma = 500$ kg/cm², have been plotted on log–log co-ordinates; the lines drawn through the points all have slopes of $\frac{1}{2}$. The effect of temperature on χ is shown in *Figure 5.6*. If the slip distance in the metal is assumed to be about half the grain diameter, then in this case $L_0 = 55$ µm. With $G = 9\cdot1 \times 10^5$ kg/cm² and $b = 2\cdot51$ Å one obtains for temperatures close to $T = 0°$K, i.e. with $a = 0$ in equation 5.20, $\chi = 20 \times 10^7$ kg/cm², which agrees well with values obtained experimentally at low temperatures.

5.3 Stress Relaxation

If a rod made of metal, most plastics or other ductile materials, is subjected to a tensile test and the test is suddenly stopped, with the material held at the final strain attained, the stress would be observed to decrease in the course of time, eventually attaining a new, stable, value. This phenomenon, known as stress relaxation at constant strain, can be understood in terms of residual plastic deformation, resulting in the smoothing out of the internal stress peaks in the material. The elastic energy of the local stress peaks is then partly dissipated by ' plastic ' work, without however resulting in an overall extension of the rod.

Provided the process occurs at temperatures sufficiently low to prevent significant contributions to atom transport by diffusion mechanisms, the internal stress peaks become reduced as a consequence of ' microscopic ' local shears. This ' dynamic ' recovery must be distinguished from high-temperature recovery and the associated high-temperature creep in which thermally activated processes contribute to significant structural changes in the material undergoing recovery.

We shall examine this relaxation in more detail, taking as a specific example a crystalline material, e.g. a metal. To simplify the calculation we shall assume that a single crystal cube of side length L has been subjected to a shear strain γ_{tot} at which it is then held. Since this total strain consists of the elastic and plastic components,

$$\gamma_{\text{tot}} = \gamma_{el} + \gamma \qquad (5.21)$$

the constancy of the total shear strain implies

$$d\gamma = - d\gamma_{el} = - d\tau/G \qquad (5.22)$$

where G is the shear modulus and τ the applied stress, which is, of course, a function of time. A plastic strain increment $d\gamma$ arises from the displacement of ρL^2 glide dislocations by a small distance dx along their respective slip planes so that, as in section 5.1

$$d\gamma = (\rho L^2)\,(b/L)\,(dx/L) = \rho.b.dx \qquad (5.23)$$

where we have considered the temperature to be low enough to enable us to assume that the density ρ of moving dislocations does not vary in the course of the relaxation experiment. From equation 5.23 one finds, directly,

$$d\gamma/dt = \rho.b.u \qquad (5.24)$$

where u is the instantaneous velocity of the dislocations.

On using equation 5.22 and the relation for the dislocation velocity (equations 4.8 and 4.11) the rate of stress relaxation is found to be given by

$$-\frac{d\tau}{dt} = G\rho b^2 v_0.\exp\left[-(Q_0 - \tau b^2 l_j)/kT\right] \qquad (5.25)$$

the 'high stress' form of equation 4.8 being used. The mean effective stress acting on the glide dislocations, appearing in the exponential term, is assumed equal to the applied shear stress; this is reasonable, for l_j is also taken to be a mean value. A more convenient way of writing equation 5.25 is

$$-\frac{d\tau}{dt} = A.\exp\left[-\frac{Q_0}{kT}\left(1 - \frac{\tau}{\tau_0}\right)\right] \qquad (5.26)$$

where

$$\tau_0 = Q_0/b^2 l_j \qquad (5.27)$$

and

$$A(\rho) = G\rho b^2 v_0 \qquad (5.28)$$

The physical significance of τ_0 may be appreciated by considering that stress relaxation could not occur at the observed, non-zero, rates at low temperatures unless as $T \to 0$ also $\tau \to \tau_0$ (equation 5.26). With our definition of τ the stress τ_0 should therefore be equal to the flow stress in shear at very low temperatures.

The dislocation density is assumed to be determined by the stress $\tau(0)$ at the time $t = 0$ at which the relaxation under constant strain begins; A is time invariant during relaxation at a given strain.

Integration of equation 5.26 yields

$$(kT/Q_0\tau_0) \exp[-(Q_0/kT)(\tau/\tau_0)] = A(t+t_0) \exp-(Q_0/kT)$$

where t_0 is a constant of integration. On taking logarithms of both sides and differentiating with respect to $\ln(t + t_0)$ one obtains

$$d\tau/d\ln(t + t_0) = -\tau_0 (kT/Q_0)$$

which yields the equation of 'logarithmic' stress relaxation

$$\tau(0) - \tau = r \ln(t + t_0) \qquad (5.29)$$

with

$$r = \tau_0 kT/Q_0 \qquad (5.30)$$

It is frequently possible to neglect t_0 after the first few seconds of relaxation, and the evaluation of r is then particularly simple. Equation 5.30 may be applied to tensile stress relaxation by replacing τ_0 by $\sigma_0/3$, as in equation 5.14. Remembering that $\sigma(0) \to \sigma_0$ as $T \to 0°K$, equation .30 can be used to evaluate Q_0 from measurements of r and τ made at low temperatures.

Experimental work on stress relaxation on copper, cobalt, nickel, iron, magnesium, and uranium, as well as on polymers such as polyethylene, on cork and other materials, have shown that the logarithmic form of stress relaxation is of wide occurrence at low temperatures. The reason for this universality lies no doubt in the generality of the assumptions embodied in equation 5.22 and in the assumed linear stress dependence of the activation energy (equation 5.25), which would be valid, at least as a first approximation, for most thermally activated processes in which the stress reduces the energy barrier.

For polycrystalline metals deformed in the 'parabolic' stage of work-hardening, values of Q_0 agree quite well with the activation energies for vacancy formation, suggesting that the non-conservative drag of intersection jogs is responsible for the frictional force on moving dislocations. The activation energy to form interstitial ions in general exceeds that for vacancy formation by a factor of about 3; interstitials do not in fact appear to form in appreciable concentrations in metals in the course of plastic deformation.

Further, equations 5.27 and 5.30 show that measurements of r provide a means for evaluating the average jog spacing l_j; this is generally found to lie in the range $100-1000b$. With the approximation $\tau = \tau_0$, valid at low temperatures, the relation for the flow stress (equation 3.19) yields with equation 5.27:

$$l/l_j \approx Gb^3/Q_0 \qquad (5.31)$$

and for most metals this ratio is approximately 5, in reasonable agreement with experiment.

5.4 Logarithmic Creep

The above discussion of stress relaxation has shown that the state of stress in a material is not determined only by the strain, but also by the temperature and the loading rate; the latter may of course be zero. One can express this formally by writing

$$d\tau = \frac{\partial \tau}{\partial t} dt + \frac{\partial \tau}{\partial T} dT + \frac{\partial \tau}{\partial \gamma} d\gamma \qquad (5.32)$$

G

In the case of isothermal tests, which we shall consider, the term $\partial\tau/\partial T$, which takes into account mainly changes of elastic constants with temperature, is zero. Further, if the material is subjected to a constant stress, $d\tau = 0$ and one obtains

$$\frac{\partial\tau}{\partial t}\ dt = -\ \frac{\partial\tau}{\partial\gamma}\ d\gamma$$

Hence the shear rate under constant stress is

$$\frac{d\gamma}{dt}\ =\ -\ [1/h(\gamma)]\ \frac{\partial\tau}{\partial t} \tag{5.33}$$

where $h(\gamma)$ is the coefficient of work-hardening which would be measured at a strain γ, at a sufficiently fast shear rate to remain unaffected by time-dependent recovery effects. The partial differential $\partial\tau/\partial t$ represents the rate of stress relaxation at constant strain. Recovery effects at low temperatures are sufficiently small to enable us to assume that $h(\gamma)$ remains at its initial value if the strain is slightly increased; this approximation is equivalent to assuming that the slope of the stress strain curve is constant over a small range of strains close to γ. Then, if both sides of equation 5.33 are multiplied by $t + t_0$ and integrated, one finds that

$$\frac{d\gamma}{d\ln(t + t_0)} = -\ \frac{1}{h(\gamma)}\ \cdot\ \frac{\partial\tau}{\partial\ln(t + t_0)} \tag{5.34}$$

or, on using equations 5.29 and 5.30:

$$\gamma - \gamma(0) = (\tau_0 kT/Q_0 h)\ \ln(t + t_0) \tag{5.35}$$

which represents the equation of 'logarithmic' creep. It may again be adapted for use with tensile deformation by using equations 5.14 and 5.15.

As $h(\gamma)$ generally decreases with increasing values of τ_0 the slope of the γ versus $\ln(t + t_0)$ curves should increase with increasing values of $\gamma(0)$. This conclusion, and the simple relation between logarithmic stress relaxation and creep, expressed by equations 5.34 and 5.35, are well borne out by experiment.

5.5 High-temperature Creep

If plastically deformed crystalline materials are held for some time at relatively high temperatures, generally in excess of about $0.4\ T_m$ (°K), where T_m is the melting temperature, they are observed to soften and the dislocation density diminishes. In polycrystals grain growth may also occur, the larger grains tending to consume the smaller ones, growing at their expense. After heavy deformation, such as can be induced by rolling, recrystallisation may set in; new grains nucleate and grow into the old, deformed ones, the stored elastic energy providing the principal driving force. High-temperature recovery, frequently termed ' annealing ' is used extensively in metal forming processes. A wire, for example, can be softened in this manner after leaving one die before being subjected to further reduction in another.

However, in high-temperature applications of materials, where the emphasis is on load bearing capacity, the enhanced susceptibility to softening and deformation arising from thermal recovery may be undesirable, for example in components of steam and gas turbines, in boilers and equipment used in the generation of electric power, or in space vehicles. The relatively slow deformations occurring in materials at elevated temperatures even under constant stresses are known as ' high-temperature creep '.

Thermal recovery tends to soften the material, but as the flow stress is thereby reduced the constant applied stress will again deform it, inducing renewed hardening. The essential feature of this dynamic balance between softening and hardening in isothermal creep may again be represented, formally, by means of equation 5.32; as the stress is constant $d\tau = 0$. However, by contrast with logarithmic creep the coefficient of work-hardening must now be con-sidered to depend not only on the strain but, as the structure of the material changes during creep, also on time.

We shall take the flow stress τ of the crystal, e.g. a metal, to be given by equation 3.19, l being the structure-sensitive parameter. This stress could be measured, in principle, if the material undergoing creep could be cooled to a tem-perature at which the deformed state is stable, without intro-ducing structural changes during the cooling. By differen-tiating equation 3.19 one finds for the rate of relaxation at constant strain

$$\frac{\partial \tau}{\partial t} = - \frac{Gb}{l^2} \cdot \frac{\partial l}{\partial t} \qquad (5.36)$$

which, with equation 5.33 yields

$$\frac{d\gamma}{dt} = \frac{Gb}{hl^2} \frac{\partial l}{\partial t} \qquad (5.37)$$

The partial differential coefficient represents the time rate of change of the characteristic dimension of the dislocation mesh which would be observed if the creep were suddenly stopped at the strain γ by removing the stress. Equations 5.36 and 5.37 show that the creep rate is determined by the ratio of the instantaneous rates of recovery and work-hardening.

When the constant stress is first applied to the material, loading conditions are akin to those encountered in a rapid tensile test, and the strain attained after a fixed, short, interval may obey a work-hardening law, such as is given by equation 5.19 for example. Following this ' initial ' rapid extension the strain rate will diminish in the course of transient or ' first stage ' creep until a dynamic equilibrium is established in which hardening and recovery occur at constant rates (*Figure 5.7*).

The creep rate then remains almost steady, and no appreciable structural changes occur in the material in this

' second stage ' or ' equilibrium ' creep, except for the gradual accumulation of localised internal damage which eventually leads to fracture. The accelerated creep, or ' stage 3 ', often observed just prior to fracture, is indicative of a progressive loss of intergranular coherency.

A detailed analysis of the cybernetics of high-temperature creep processes is still outstanding. However, considerable insight into creep mechanisms can be obtained by considering certain features of equation 5.37, making the

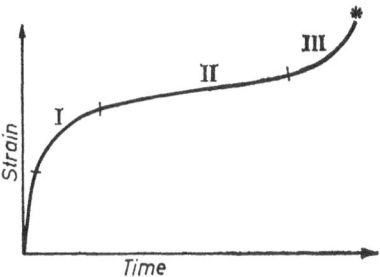

Figure 5.7. The three characteristic stages of high-temperature creep, following after the initial, rapid extension

assumption that variations in the dislocation mesh dimensions occur at a rate determined by the dislocation velocity, such that equation 5.37

$$\frac{\partial l}{\partial t} = cu \tag{5.38}$$

In equilibrium creep, which we shall now consider, c will be taken to be a constant of order unity. One then obtains from equations 5.37, 5.38 and 4.8, using the high-stress approximation, the equation for the equilibrium tensile creep rate:

$$\frac{d\epsilon}{dt} = \frac{Gb^2 c \nu_0}{3hl^2} \cdot \exp\left[-(Q_0 - \tfrac{1}{3}\sigma b^2 lj)/kT\right] \tag{5.39}$$

93

where equations 5.14 and 5.15 were used to replace shear stresses and strains by equivalent tensile parameters. The values of h, l and l_j are assumed to be those characteristic of the equilibrium creep. If reasonable assumptions are made, for example $h = G/200$ and $l = 10^{-4}$ cm, also taking $c = 1$, the pre-exponential fraction in equation 5.39 is found to be about 4×10^5 per second for copper. This agrees quite well with measured values obtained at about 700°C, which is equivalent to about $0 \cdot 7 \; T_m$ (°K) in this metal. In this case, and generally above about $0 \cdot 5 \; T_m$, Q_0 is not identifiable with the activation energy for vacancy formation, as in logarithmic creep; it tends to be close to the activation energy of self-diffusion. This observation can be explained if it is considered that diffusional processes can occur readily at high temperatures, so that formation of a vacancy at a jog would not necessarily allow the jog to climb unless either the dislocation moves so fast that the jog is displaced from the vacancy before the latter can jump back to it, or if the vacancy, once formed, jumps away from the jog.

The first alternative is improbable in conventional high-temperature creep tests, as the dislocations would not move fast enough. In the second case the chain of events facilitating non-conservative jog displacements consists of the formation of a vacancy, followed by its diffusion away from the jog. The activation energy would then comprise the energies of vacancy formation and migration; their sum is in general close to the activation energy of self-diffusion.

5.6 Creep Fracture

Vacancies transported into the material by the dislocations in the course of creep will tend to segregate in the relatively open structure close to the grain boundaries or at local stress fields due to impurities in solid solution and, also, around inclusions. Such heterogeneous precipitation, leading to the aggregation of vacancies into small holes can be a source of fracture. If, for example, the damage attains a critical state in which small pores coalesce to form cracks the metal may fail rapidly. On the basis of the assumed accumulation of damage up to a critical amount a high creep rate would be expected to lead to a high rate of this process and hence to a relatively short creep life of the specimen, and vice versa. Also, since the second stage creep would in general prevail during most of the time during which the specimen is under load, one would expect to find the time to fracture t_f to be approximately inversely proportional to the equilibrium creep rate, i.e.

$$\left(\frac{d\epsilon}{dt} \right)_{II} \approx m/t_f \qquad (5.40)$$

where m is a dimensionless constant numerically close to the strain at fracture. The equation is reasonably well obeyed by pure metals and solid solutions, but not by highly heterogeneous materials.

In practice materials may be subjected to creep in use over many years; criteria of technical creep strength must then be specified in relation to the function and expected creep life of the material or structure in which it is embodied. For example, an alloy may be specified for a certain application such that under stated stresses and at a certain temperature it should not creep faster in tension than $0 \cdot 01$ per cent per hour when the creep strain has attained $0 \cdot 2$ per cent. Alternatively it may be required that after a fixed period under load at a given temperature, e.g. after 45 h, the specified stress should not lead to strains greater than $0 \cdot 2$ per cent, and the creep rate should not exceed $0 \cdot 01$ per cent per hour after, say, 30 h on test.

5.7 Point Defect Concentrations

In view of the importance of point defects, particularly vacancies and vacancy aggregates, in plasticity and fracture we shall attempt to obtain an estimate of the vacancy concentrations arising in the course of work-hardening. We shall again use face-centred cubic metals as our example, and make the simplifying assumption that all the vacancies are formed in rows by non-conservatively moving jogs, one vacancy being formed per displacement b of the dislocation.

According to equation 5.31 the spacing between non-conservatively moving jogs l_j may be expressed in terms of the dislocation mesh l by

$$l \approx 5 \, l_j \tag{5.41}$$

which, on writing $l = \rho^{\frac{1}{2}}$, yields with equation 5.3:

$$L = 5 \, (c_1 \, c_2)^{\frac{1}{2}} \, l_j \tag{5.42}$$

Now the number of jogs per unit length of dislocation will increase as the dislocation migrates over the slip distance L; they will first be relatively far apart, but the spacing will gradually diminish. The mean spacing between jogs on a given dislocation, considered over the period of its passage, will therefore be larger than the final value l_j referred to in equation 5.42, say $c_j \, l_j$, where we shall assume $1 < c_j < 10$. The total number of vacancies generated per centimetre of dislocation line in the course of slip will be

$$N = (L/b) \, (1/c_j l_j) \tag{5.43}$$

and with the previous estimates of $c_1 = 100$ and $c_2 = 1000$, also taking $c_j = 5$, we obtain from equations 5.42 and 5.43:

$$N \approx 300/b \tag{5.44}$$

For most metals this yields

$$N \approx 10^{10} \text{ cm}^{-1} \tag{5.45}$$

Measurements of the excess resistivity due to point defects in deformed metals at low temperatures suggest that this estimate may be too high by a factor of about 10. It should be noted, however, that a uniform dispersion of vacancies is generally assumed in calculating vacancy concentrations from excess resistivity data, and this may lead to an under-estimate of N.

Thus a heavily cold-worked metal with a dislocation density of 10^{12} cm^{-2} should contain about 10^{21} vacancies per cm^3 due to cold-work. If these were condensed they would ' fill ' a cube of about $2 \cdot 5$ mm length of side. Their presence should lead to a detectable reduction in the density of the material. Such density changes are observed, in part however they also arise from dislocations. On heating the metal to temperatures where vacancies become mobile, e.g. about 400°C in copper, self-diffusion should lead to a rapid drop of the vacancy concentration to the rather small equilibrium concentration; its magnitude at temperatures at which vacancies can diffuse at significant rates is given approximately by exp $(-Q_0/kT)$, where Q_0 is the energy of formation of a vacancy. Evidence that vacancies also accumulate in ionic crystals on deformation at rates comparable to those here estimated has been obtained.

6

FRACTURE AND FATIGUE

6.1 Brittle Fracture

Two principal types of fracture are generally distinguished. Fracture preceded by a significant amount of plastic deformation is known as ' ductile ', otherwise it is ' brittle '. The latter, which we shall consider first, occurs when plastic flow is inhibited, be it by the effective locking of dislocations by precipitates or elements in solid solution or by the pre-existence or formation of cracks and imperfections acting as local stress raisers in the material. All materials can be embrittled if the temperature is lowered sufficiently. Glass, sealing wax, germanium, silicon and other materials, though ductile at temperatures close to their melting points, are brittle at ordinary temperatures; as was previously mentioned, rubber cooled in liquid nitrogen may be shattered with a hammer.

In most materials the brittle strength, defined as the maximum tensile stress withstood without the occurrence of brittle fracture, is low compared with the ideal strength the fault-free material would be expected to exhibit. The source of brittle fracture has therefore to be sought in the presence of structural defects.

Most of the theoretical work on brittle fracture has in fact been developed on the basis of the ' crack ' concept introduced by A. A. Griffith in 1924. He considered that brittle materials contained microscopic cracks which could extend their lengths abruptly if the tensile stress across them exceeded a definite critical value. This spread of cracks was assumed to lead to brittle fracture.

As long ago as 1913, C. E. Inglis had calculated the stress distribution in an elastic plate containing an elliptic crack of length $2c$, subjected to a tensile stress σ (*Figure 6.1*). He found that the tensile stress is increased at the ends of the crack by the ' notch factor ' to $2(c/R)^{\frac{1}{2}}\sigma$, R being the radius of curvature at the ends of the long axis of the ellipse. The elastic energy stored in the plate due to the presence of the crack

$$W(\sigma) = \pi c^2(\sigma^2/E) \tag{6.1}$$

increases with the crack length and the applied stress. Griffith considered that the crack would spread by extending

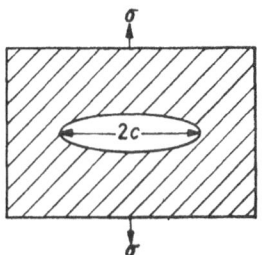

Figure 6.1. An elliptical Griffith crack subjected to a tensile stress σ

its length, and hence the fracture surfaces, when the stress attained a critical value σ_{B}, the brittle strength. Its value depended on the specific surface energy s of the crack faces, the latter requiring an energy

$$W(s) = 4cs \tag{6.2}$$

for their formation. The crack would spread when $\partial W(\sigma)/\partial c$ just exceeded $\partial W(s)/\partial c$, and this criterion yields

$$\sigma_{\text{B}} = (2sE/\pi c)^{\frac{1}{2}} \tag{6.3}$$

An estimate of the theoretical cohesive strength can be obtained from equation 6.3 on replacing c by the smallest physically meaningful length, an interatomic spacing in the preferred cleavage plane of the material. In freshly drawn fine glass rods σ_B has been found to be close to the theoretical value, but it diminished by a factor of 30 or more in the course of several hours, suggesting that cracks exceeding 1000 interatomic spacings in length must have formed in that time. Direct evidence in support of this inference was obtained by microscopy of the surface. In brittle metals the relation

$$\sigma_B \propto L_0^{-\frac{1}{2}}$$

is observed, where L_0 may be the mean grain radius or diameter. This suggests that the damaging cracks initially extend across the grains or run along large areas of certain grain boundaries.

The weakening effect of stress raisers, such as notches on the surface of the material, or the presence of sharp inclusions within it, are well known. A classical example is provided by the internal notches due to graphite flakes in cast-irons. The flakes embrittle the irons in tension; in structural applications cast-irons are therefore more usefully employed under compressive loads. Their brittle strength and toughness can, however, be increased appreciably if the graphite is allowed to form in spheroidal rather than flaky form. This can be achieved by alloying the melt, for example with magnesium; the resulting product is known as ' SG ' (spheroidal graphite) iron.

6.2 Ductile Fracture

In single crystals deforming by slip on one preferred slip system, fracture may occur through ' slipping off'; one part of the crystal shears away entirely from another. Otherwise ductile fracture can occur as a result of the spread of small fissures, developing in ' bands ' of closely spaced slip planes which have been particularly active in the deformation. The fissures form at points of high dislocation density where ' debris ' left by dislocation interaction, and intense short-range stresses at local dislocation complexes, facilitate decohesion. Cracks may also form at inclusions and other flaws in the course of deformation.

The spreading of cracks from the surface to the interior of stressed thin metal foils has been observed by electron transmission microscopy. Propagation is generally ahead of the tip, where the stresses are highest due to the notch effect. A profusion of dislocations may be seen moving into or away from the tip, transporting material or ' vacancies'. Plastic deformation thus clearly facilitates extension of the damage.

The damage eventually leading to fracture appears to accumulate in the course of deformation. In copper for example the length of individual cracks and the number of cracks increase approximately linearly with deformation, but coalescence of cracks in the late stages of deformation does lead to a rather higher growth rate of the most ' dangerous ' cracks. One would therefore expect a relation of the form

$$c = c_0 \, (1 \, + \, a\epsilon^{1+\beta}) \qquad (6.4)$$

to describe the crack growth kinetics reasonably well, $2c_0$ being the crack length in the unstrained material, and a and β positive constants appropriate under conditions of tensile deformation. If the stress across the cracks is taken to be proportional to the applied tensile stress σ then, writing $\sigma = \sigma_f$ at fracture, equation 6.3 becomes

$$\sigma_{\mathrm{B}} = C\sigma_f = (2sE/\pi c_f)^{\frac{1}{2}} \qquad (6.5)$$

where C is a constant of proportionality and c_f refers to the half length of the cracks at fracture. From equations 6.4 and 6.5 one obtains

$$\sigma_f^2 = K/(1 \, + \, a\epsilon^{1+\beta}) \qquad (6.6)$$

where K is a constant which would not be expected to be significantly temperature dependent. If, as in the case of the cobalt specimen (*Figure 5.4*) the work-hardening can be described very well by the relation

$$\sigma^2 = \chi(T) \cdot \epsilon \qquad (6.7)$$

which also holds when $\sigma = \sigma_f$, then by eliminating the strain between equations 6.6 and 6.7, one finds that

$$\sigma_f^2 = K/[1 + a\,(\sigma_f^2/\chi)^{1+\beta}] \qquad (6.8)$$

Now, as c_f would be expected to be much larger than c_0 (equation 6.4), the term $a(\sigma_f^2/\chi)^{1+\beta}$ should also be

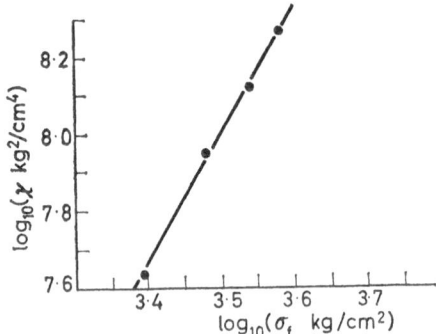

Figure 6.2. *Log–log representation of the χ versus σ_f relation for cobalt. The slope of the curve yields $m = 0 \cdot 30$*

appreciably larger than 1, so that one should have approximately

$$\sigma_f^2 \propto (\chi/\sigma_f^2)^{1+\beta} \qquad (6.9)$$

or where $m = (1 + \beta)/(4 + 2\beta)$

$$\sigma_f \propto [\chi(T)]^m \qquad (6.10)$$

The limiting values of m are therefore $\frac{1}{4}$ and $\frac{1}{2}$, i.e. for $\beta = 0$ or $\beta \to \infty$. *Figure 6.2* shows that the hypotheses made in deriving this result appear to be reasonable; the value obtained for cobalt is $0 \cdot 30$, corresponding to $\beta = 1$. For pure copper deformed in the same temperature range one obtains $m = 0.40$, which is also well within the permitted limits of m. Although the uniqueness of the preceding model has not been established, the results are consistent with the view that fissures and the simultaneous increase of the applied stress eventually lead to the rapid spreading of cracks and consequent fracture.

If the specimen were deformed in shear, for example by twisting a rod, wire or tube, an equation corresponding to

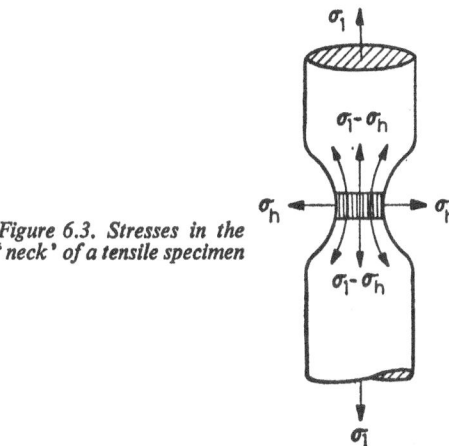

Figure 6.3. Stresses in the 'neck' of a tensile specimen

6.4 would still apply, but in the absence of applied tensile stresses the local internal tensile stresses would be expected to be relatively small, and large shear strains might be anticipated to precede failure. This is in agreement with experience; shear strains of 300 per cent or more are readily obtained by twisting metal wires such as silver or aluminium at ordinary temperatures, and these strains are greatly in excess of the corresponding values of $3\epsilon_f$ (equation 5.15), where ϵ_f is the strain at fracture in tensile specimens of the same material.

103

Also, if high hydrostatic stresses are superimposed upon the applied stress in a tensile test ductility is found to be enhanced. This can be explained if cracks are inhibited from opening up by the hydrostatic compression. Conversely, ductility is in general lowered if the specimen is notched. The origin of this effect can be seen on considering a tensile specimen which has a ' neck ', the latter being equivalent to a circumferential notch (*Figure 6.3*). The forces due to the applied stress σ_1 give rise to a transverse component in the neck region, so that the stress system can be resolved into an axial component $\sigma_1 - \sigma_h$ and a hydrostatic tension σ_h. If the flow stress in tension of a notch-free specimen is σ then for the necked sample

$$\sigma = \sigma_1 - \sigma_h$$

Consequently σ_1 must be increased to $\sigma + \sigma_h$ before flow will set in around the neck. The hydrostatic component of the stress will assist in the opening up of cracks, so that the material is embrittled at the same time as the ' apparent ' flow stress σ_1 is increased. In plastic materials σ_h cannot be increased indefinitely, e.g. by making the notch very sharp; it may be shown that σ_1 can attain a value of at most 3σ.

6.3 Fatigue

Fatigue is a mode of fracture which occurs when materials are subjected to alternating loads over prolonged periods at stress levels which would not lead to failure under static loading. It is a universal phenomenon, observed in most solids, and a common specific fracture mechanism cannot therefore exist, except in so far as cyclic loading leads to a continuous accumulation of damage which, as in the case of static fracture, eventually results in rupture.

Although the problem was studied by Albert almost 130 years ago, general awareness of the fact that the tensile strength is not a measure of the resistance of materials to fracture under all conditions encountered in practice was aroused only early in the 20th century as a consequence of the mysterious failure of some cast-iron bridges.

The essential features of fatigue, and measures which have to be taken to prevent its incidence, are nowadays quite well known, yet despite such powerful methods of investigation as electron transmission microscopy and X-ray micro-focus techniques it is likely to remain a quarry for research for some time to come.

Fatigue is a serious problem with metals for these are most widely used in dynamically loaded structures and machines. We shall therefore confine our attention to

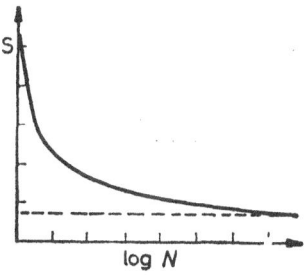

Figure 6.4. The relation between the peak stress S and the number N of cycles to fracture in alternating tension–compression with zero mean stress

metals and alloys, without thereby making a significant sacrifice of generality.

When a metal tensile specimen is subjected to tension–compression cycles with zero mean stress one finds in general that the relation between the peak stress S and the number of cycles to failure is of the form shown in *Figure 6.4,* known as the S–N or Wöhler curve. Semi-logarithmic co-ordinates are used. In practice the curve rarely becomes truly horizontal, so that there is no ' safe range ' of stresses, and fracture would ensue after a sufficiently large number of stress cycles. In certain alloys which have a well defined yield point, for example in plain carbon steels, the curve may have an asymptote, shown as a dashed line in the figure.

The stress corresponding to its intersection with the S-axis is then the ' true ' fatigue limit.

Determining the true fatigue limit is unnecessary and uneconomical in practical applications, and a ' technical ' fatigue limit is determined instead. This may be specified as the stress at which the material will endure a given number of cycles, for example 10 million, without fracture. The number of cycles leading to fracture at a given stress is often referred to as the ' fatigue strength ' or ' endurance ' at that stress.

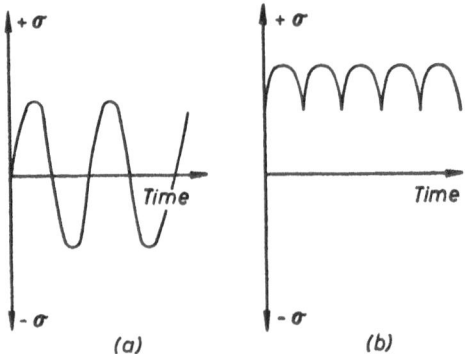

Figure 6.5. Fatigue cycles with (a) zero and (b) positive mean stress

If the loading is not simple harmonic, with zero mean stress, but a constant tensile or compressive stress is super-imposed upon the oscillatory component, then a positive or negative mean stress results (*Figure 6.5*). For a given peak stress symmetric stressing, with zero mean stress, will in general lead to a longer fatigue life than the use of asym-metric cycles with the same peak stress. This is mainly due to the more complete healing of fatigue damage in the first case, for microscopic local stress peaks developed in, say, a positive cycle will be more nearly annulled by a complete reversal of the applied stress than by a partial reversal only. The S–N curve for asymmetric stressing will therefore lie below that for simple harmonic loading.

In the early stages of fatigue testing, specimens will generally evolve an appreciable amount of heat due to plastic work, and they may work-harden somewhat. Later fissures develop and spread from slip bands at the surface, eventually leading to fracture. The surface is a preferential seat of damage initiation, for slip is less confined there than in the interior, and corrosive effects may also assist in the degradation of the structure at the surface.

The fatigue strength of metals can often be enhanced by treatments which render the surface more resistant to deformation; fracture then tends to start at the interface between the hard surface layer and the softer core. Stress raisers, such as sharp notches, corners, key-ways, rivet holes and scratches can lead to an appreciable lowering of the fatigue strength of metal components. Good surface finish and corrosion protection are frequently desirable to enhance fatigue resistance. Fatigue is essentially a low-temperature problem; at temperatures relatively high with respect to the melting point, fracture (and hence specimen life) are governed by creep.

The mechanism of metal fatigue is complex, and not fully understood. Available evidence shows, however, that damage accumulates gradually, as in ductile fracture and creep, until the specimen or component is weakened to the point where ordinary processes of fracture terminate its useful life. This view finds support in the observation that fatigue life is not significantly affected by the frequency of the loading cycles over wide ranges of frequency, but depends mainly on the total number of stress reversals.

Formation of cracks and cavities in the course of plastic deformation is favoured by extensive interaction between dislocations. In fact, in pure face-centred cubic metals the true fatigue limit seems to coincide with the stress at which the third, parabolic, stage of work-hardening begins. This stage, as was pointed out before, is accompanied by pronounced dynamic recovery resulting from dislocation interactions.

The gradual accumulation of damage is also clearly indicated by the observation that if a specimen is annealed after it has spent part of its expected fatigue life, the latter cannot in general be extended, unless the specimen is allowed to recrystallise by annealing at temperatures relatively high with respect to the melting point. A further similarity between fatigue and fracture lies in the temperature dependence of the technical fatigue limit and the fracture strength, which seems to be approximately the same in both cases. In commercial alloys the technical fatigue limit generally lies between $0 \cdot 3$ and $0 \cdot 5$ of the ultimate tensile stress. This range can however serve only as an approximate guide, for with materials of relatively complex structures it is difficult to make simple, precise, generalisations.

Fracture surfaces of fatigued metals generally show a smooth, lustrous, region, due to the polishing and abrading effects arising from attrition at fissures. The remaining parts of the fracture surface, over which failure occurred through weakening of the specimen by the reduction of its load bearing cross-section by surface cracks and fissures, may look duller and coarser, as it is essentially due to static fracture.

Incipient fatigue can sometimes be detected, even though fissures are not yet detectable, by a pronounced increase in the damping capacity or ' internal friction ' of specimens cut from the fatigued piece. This enhancement of the damping capacity is largely a consequence of the dissipation of part of the elastic energy of the vibrating or oscillating specimen by friction at crack interfaces. The method is however destructive; it is not possible in most cases to cut test specimens from operating structures or machines. It is therefore only useful as a research tool. A more detailed discussion of the principles involved in the study of internal friction will be given in the next chapter.

7

RHEOLOGICAL MODELS AND DAMPING CAPACITY

7.1 Damping Capacity

Dissipation of elastic energy, which manifests itself, for example, in the gradual decay of oscillations 'of a torsion pendulum and of the vibration amplitude of a tuning fork, even if air friction is practically eliminated altogether, is invariably observed in all materials. Energy may be dissipated in the formation of, or through abrasion at, cracks, by the displacement of ions from their mean positions in the lattice due to the to-and-fro movement of dislocations, by work expended in pulling dislocations away from point defect or impurity pinning points, and by a variety of other thermal, electrical, magnetic and diffusion processes; all have the common feature however of absorbing some of the elastic energy of oscillation or vibration stored in the material.

The simplest model of a damped solid consists of an ideally 'Hookean' element such as a longitudinally vibrating elastic rod, coupled in parallel with a viscous element or 'body'; a spring and a dashpot are in general used as the respective symbols, as shown in *Figure 7.1*.

For tensile deformation Young's modulus E and Trouton's coefficient of viscous traction ζ are used, while for deformations by shear they are replaced by the shear modulus G and the coefficient of viscosity η. With a Newtonian liquid in the dashpot ζ and η are stress and time independent. On taking σ_ζ to be the component of the

110

applied tensile stress σ acting on the dashpot only, one has

$$\sigma_\zeta = \zeta \,.\, \mathrm{d}\epsilon/\mathrm{d}t \tag{7.1}$$

where $\mathrm{d}\epsilon/\mathrm{d}t$ is the tensile strain rate of both spring and dashpot. The parallel-connected combination of spring and dashpot shown in *Figure 7.1* is known as the Voigt or Kelvin body. It is a solid, for it will not suffer permanent

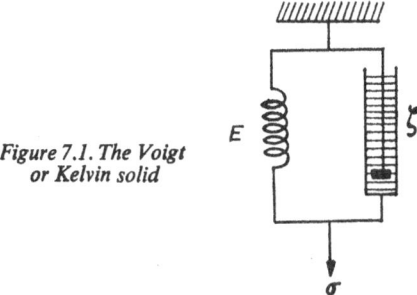

Figure 7.1. The Voigt or Kelvin solid

deformation, always regaining its initial shape when the stress σ is removed. In the shear representation the equation equivalent to 7.1 is

$$\tau_\eta = \eta \,.\, \mathrm{d}\gamma/\mathrm{d}t \tag{7.2}$$

while the respective equations for the Hookean element are

$$\sigma_E = E \,.\, \epsilon \tag{7.3}$$

and

$$\tau_G = G.\gamma \tag{7.4}$$

Again, as in equations 5.14 and 5.15, we may assume that σ_E is a constant multiple of τ_G, etc., and that, similarly, ϵ is proportional to γ. The relations

$$\frac{\sigma_\zeta}{\sigma_E} = \frac{\zeta}{E} \,\cdot\, \frac{\mathrm{d}\epsilon/\epsilon}{\mathrm{d}t}$$

$$\frac{\tau_\eta}{\tau_G} = \frac{\eta}{G} \cdot \frac{d\gamma/\gamma}{dt}$$

obtained from equations 7.1 to 7.4, then imply that

$$\zeta/E = \eta/G$$

and since, for an incompressible material, $E/G = 3$ (equation 1.26), we obtain the frequently quoted relation

$$\zeta = 3\eta \qquad (7.5)$$

between Trouton's coefficient of viscous traction and the coefficient of viscosity.

If one sets the body into oscillation, by releasing a mass which has first been suspended from it, one can evaluate the energy dissipated per cycle by the dashpot as follows. Assuming that the damping is sufficiently small to enable us to write for the strain over the period of one cycle

$$\epsilon = \epsilon_0 \cos \omega t \qquad (7.6)$$

where ϵ_0 is taken to be constant over that particular cycle, i.e. for $0 \leqslant t \leqslant 2\pi/\omega$, where ω is the frequency of oscillation in radians per unit of time, the energy dissipated is

$$\Delta w = \int_{t=0}^{2\pi/\omega} \sigma_\zeta \frac{d\epsilon}{dt} \, dt \qquad (7.7)$$

and this yields with equations 7.1 and 7.6:

$$\Delta w = \pi \zeta \omega \epsilon_0^2$$

With the assumed constancy of ϵ_0 over the cycle considered the elastic energy stored per unit volume is, by equation 1.20:

$$w = \tfrac{1}{2} E \epsilon_0^2$$

and one has

$$\Delta w/w = 2\pi \, (\zeta/E) \, \omega = 2\pi \, (\eta/G) \, \omega \qquad (7.8)$$

which is independent of ϵ_0. The ratio η/G is generally referred to as the relaxation time of the system.

Several measures of the damping capacity are in use. Of these the 'logarithmic decrement' and 'Q' are related to $\Delta w/w$ as shown in equation 7.9

$$\tfrac{1}{2} \Delta w/w = \pi/Q = \text{log. dec.} \qquad (7.9)$$

The assumption that the viscosity is Newtonian, used in the above discussion of the Voigt–Kelvin model, is generally not restrictive in considerations of damping capacity, for the latter is in most cases measured at sufficiently small stresses to justify the assumption of a linear relation between stress and strain rate, implied by the use of Newtonian viscosity. It can be seen, for example, that the velocity u of dislocations, and hence the creep rate (equations 4.8 and 5.24) increase linearly with the stress if the latter is sufficiently small, and the dislocation density is constant, as could be the case at temperatures which are not too high in relation to the melting point. Under these conditions the flow of the material will therefore appear to be Newtonian.

7.2 The Standard Linear Solid

Solutions of polymers and other materials of high molecular weight often behave in a viscoelastic manner, i.e. like elastic solids when subjected to rapid loading, for example bouncing like a rubber ball, yet flowing like a viscous liquid, with permanent loss of shape, when subjected to static or slowly varying stresses. The simplest model with such behaviour is the Maxwell liquid, shown in *Figure 7.2*. It exemplifies creep as well as stress relaxation. The total strain ϵ now consists of the component $\epsilon_E = \sigma/E$ due to the elastic element, and a contribution made by the dashpot, defined by

$$\sigma = \zeta \, . \, \mathrm{d}\epsilon_\zeta/\mathrm{d}t$$

so that

$$\mathrm{d}\epsilon/\mathrm{d}t = \mathrm{d} \, (\epsilon_E + \epsilon_\zeta)/\mathrm{d}t$$

i.e.

$$\mathrm{d}\epsilon/\mathrm{d}t = E^{-1}\mathrm{d}\sigma/\mathrm{d}t + \sigma/\zeta \qquad (7.10)$$

113

which is the differential equation describing the behaviour of the model.

If the stress is maintained constant, the first term on the right-hand side of equation 7.10 is zero, and the material flows, or creeps, at a constant tensile strain rate σ/ζ. At a constant strain ϵ_0, the strain rate $d\epsilon/dt$ is zero, and

$$d\sigma/dt = -\sigma (E/\zeta) \qquad (7.11)$$

The stress then relaxes from its initial value in accordance with the exponential relation

$$\sigma(t) = \sigma(0) \cdot \exp(-Et/\zeta) \qquad (7.12)$$

It can be seen that after a period equal to the relaxation time ζ/E the stress has dropped to $1/e$ of its initial value.

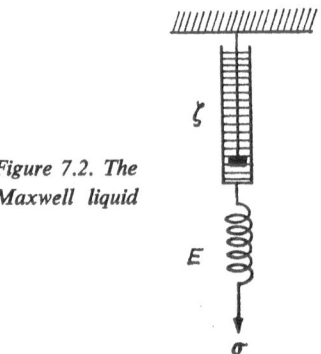

Figure 7.2. The Maxwell liquid

The simple rheological models so far discussed are rarely adequate to describe the behaviour of real materials even at low stresses, and more complex bodies consisting of combinations of Voigt–Kelvin and Maxwell bodies are then employed. A model which has been used extensively in the study of damping phenomena is the 'Standard Linear Solid' (*Figure 7.3*). It can account for the commonly made observation that the total reversible deformation attained by, say, a wire subjected to a constant stress does not occur in its entirety at the instant of loading; the wire continues to extend slightly at a diminishing rate for some time after the load has been applied. Neither the Voigt–Kelvin nor the Maxwell bodies can explain this behaviour adequately. This reversible but time-dependent extension, following the elastic deformation, has been termed by C. Zener ' anelastic ' to distinguish it from immediate, ideal, elastic response on the one hand and from irreversible, non-elastic, deformation on the other.

If the loading rate is so high that the plunger in the dashpot has moved only very little by the time the full load is applied the dashpot may be regarded as rigid during that period, and by considering that the stresses acting on both spring elements give rise to the same strain in each, one finds that the effective ' unrelaxed ' elastic modulus is E. The subsequent movement of the dashpot under load eventually renders the spring with modulus ΔE ineffective, and the final, relaxed modulus is $E - \Delta E$. The increment ΔE is known as the ' modulus defect '.

By a procedure similar to that adopted above, one finds for the standard linear solid the equation

$$\sigma + a_\sigma \frac{d\sigma}{dt} = (E - \Delta E) \left(\epsilon + a_\epsilon \frac{d\epsilon}{dt} \right) \qquad (7.13)$$

where the relaxation time at constant strain a_σ and the retardation time at constants stress a_ϵ are given by

$$a_\sigma = \zeta/\Delta E; \quad a_\epsilon = [E/(E - \Delta E)] \, a_\sigma \qquad (7.14)$$

In practice $\Delta E \ll E$, so that the relaxation and retardation times will be numerically almost equal.

If the solid is subjected to a periodic stress $\sigma = \sigma_0 \cos\omega t$, one finds that the damping capacity is now given by

$$Q^{-1} = \frac{\Delta E}{E} \left(\frac{\omega a}{1 + \omega^2 a^2} \right) \qquad (7.15)$$

where the relaxation and retardation times have both been equated to a, i.e. E has been written for $E - \Delta E$. On plotting Q^{-1} as a function of the frequency a curve with a maximum at a frequency ω_0 is obtained (*Figure 7.3*), where

$$\omega_0 a = 1 \qquad (7.16)$$

The corresponding maximum of Q^{-1} equal to $\tfrac{1}{2}(\Delta E/E)$, is known as the ' relaxation strength '; it enables us to

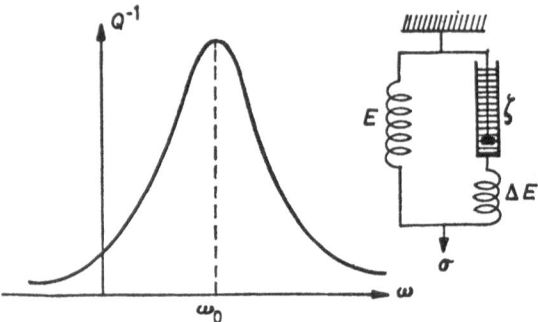

Figure 7.3. Frequency dependence of the damping capacity of the standard linear solid

determine ΔE. Then, with a obtained experimentally with the aid of equation 7.16, ζ may be evaluated (equation 7.14).

Equation 7.16 is particularly useful, for it can often be used to evaluate the activation energy of the process responsible for the damping, and it can thus assist in the identification of the atomic mechanism. This can be seen as follows.

We assume, as an example, that the damping under investigation arises from ' plastic ' work due to a limited amount of slip associated with small dislocation displacements about their equilibrium positions. If the dislocation

motion is described by equations 5.24 and 4.8, one obtains
$$d\gamma/dt = (1/\eta)\tau \tag{7.17}$$
with
$$\eta = \eta_0 \exp(Q_0/kT) \tag{7.18}$$
and
$$1/\eta_0 = 2\rho \nu_0 b^4 l_j/kT \tag{7.19}$$
where equation 4.11 was used to yield the specific form of equation 7.19. The temperature dependence of η_0 will in general be small compared with that of η, and if it is neglected, the relaxation time $\eta/\Delta G$ can be written in the form
$$a = a_0 \exp(Q_0/kT) \tag{7.20}$$
where a_0 is a temperature-independent constant.

If now the temperature of the specimen is changed from T_1 (°K) to, say, a lower one T_2, the frequency of the damping peak will decrease from ω_{01} to ω_{02}; it is readily seen from equation 7.20 that in fact
$$\ln \frac{\omega_{01}}{\omega_{02}} = \frac{Q_0}{k} \left[\frac{1}{T_2} - \frac{1}{T_1} \right] \tag{7.21}$$
One can therefore determine Q_0 by measuring the frequency of the damping maxima at two temperatures. The method has been used extensively in studies of internal friction phenomena in solids.

7.3 Electrical Analogues

The mathematical treatment of rheological bodies consisting of assemblies of springs and dashpots is analogous to that of electrical $L-C-R$ networks. For example, equation 7.12 is of the same form as that describing the decay of the voltage $V(t)$ across a condenser of capacity C which is being discharged through a resistance R:
$$V(t) = V(0) \cdot \exp(- C^{-1} t/R)$$
Stress, strain and strain rate correspond to potential, charge and current respectively, while inductance, capacitance and resistance have their counterparts in mass, the

117

inverse of an elastic modulus (a compliance), and the coefficient of viscosity respectively. The electrical analogue of the Voigt–Kelvin body (*Figure 7.1*) consists of a capacitance in *series* with an ohmic resistance, that of the Maxwell liquid (*Figure 7.2*) comprises a capacitance connected in *parallel* with a resistance.

With constant values of the network components the current flow is represented by linear differential or integral equations; analogous equations are obtained for spring and dashpot systems with constant rheological parameters. This constancy of the 'network' elements cannot be assumed with work-hardening materials, except in studies of their anelastic behaviour over narrow ranges of the variables. The structural changes which accompany plastic deformation, such as increases in the dislocation density and hence hardness, in general preclude the use of such differential equations of state even at constant temperature.

In the case of the 'linear', 'viscoelastic' or 'Boltzmann' bodies, consisting of constant m, ζ and E elements, it is possible to deduce useful equations relating to creep, relaxation and deformation under periodic or aperiodic stresses by methods of the operational calculus devised by O. Heaviside towards the end of the 19th century, mainly with the aim of analysing the response of electrical networks to transients. There are numerous textbooks on the subject, and we shall confine ourselves here to a few results of special relevance to the topic under discussion.

Now, the Carson transform $f^*(p)$ of a function $f(t)$ is defined by

$$f^*(p) = p \int_0^\infty e^{-pt} f(t) \, . \, dt; \quad t \geqslant 0 \qquad (7.22)$$

The integral, which represents the Laplace transform of $f(t)$ has been tabulated for a large number of functions.

In shorthand notation equation 7.22 is written

$$f(t) \supset f^*(p) \tag{7.23}$$

We are particularly interested in the result

$$\frac{d}{dt} \int_0^t f(u) \cdot g(t-u) \cdot du \supset f^*(p) \cdot g^*(p) \tag{7.24}$$

which we give without proof, and which we shall utilise below.

7.4 Boltzmann Superposition

If a constant stress is applied to the Voigt–Kelvin body (*Figure 7.1*), then from the condition $\sigma(0) = \sigma_E + \sigma_\xi$ one obtains, on using equation 7.3, and 7.4:

$$\epsilon(t) = \sigma(0) \cdot [1 - \exp(-Et/\zeta)]/E \tag{7.25}$$

and for a system of coupled linear bodies one can write, similarly,

$$\epsilon(t) = \sigma(0) \cdot f(t) \tag{7.26}$$

where $f(t)$ is the flow or creep function, and is taken to be zero for negative arguments.

Now suppose that at a time θ after the application of the stress a further increment $\Delta\sigma$ is applied. One then has

$$\epsilon(t) = \sigma(0) \cdot f(t) + \Delta\sigma \cdot f(t - \theta)$$

and since the initial extension must necessarily occur in a finite time, it can also be considered for the sake of generality to consist of a series of consecutive stress increments. Hence for any number of changes of stress

$$\epsilon(t) = \sum_{\theta=0}^{t} f(t-\theta) \cdot \Delta\sigma(\theta)$$

On replacing the summation by integration, also writing $t - \theta = u$, one obtains the superposition integral

$$\epsilon(t) = \int_0^t f(u) \frac{d\sigma(t-u)}{d(t-u)} du$$

As $\mathrm{d}(t - u)/\mathrm{d}t = 1$, and $f(u)$ is not a function of time, this may be written

$$\epsilon(t) = \frac{\mathrm{d}}{\mathrm{d}t} \int_0^t f(u) \cdot \sigma(t-u) \cdot \mathrm{d}u \qquad (7.27)$$

We shall use this result in establishing a relation between creep and relaxation for linear bodies.

We first take the Carson transform of $\epsilon(t)$, which, on comparing equations 7.24 and 7.28, we find to be

$$\epsilon^*(p) = f^*(p).\sigma^*(p) \qquad (7.28)$$

Similarly if, by analogy with equation 7.26, one writes

$$\sigma(t) = \epsilon(0) \cdot r(t) \qquad (7.29)$$

where $r(t)$ is the relaxation function, then

$$\sigma^*(p) = r^*(p) \cdot \epsilon^*(p) \qquad (7.30)$$

It follows from equations 7.28 and 7.30 that

$$f^*(p) = 1/r^*(p) \qquad (7.31)$$

so that if $f(t)$ is known $r^*(p)$ may be found and, consequently, $r(t)$ can be determined. Also, of course, if the relaxation function is known the creep function may be deduced.

For the Maxwell liquid, for example, noting that $\epsilon(0) = \sigma(0)/E$, we have from equations 7.12 and 7.29:

$$r(t) = E \exp(- Et/\zeta) \qquad (7.32)$$

This yields

$$r^*(p) = Ep/[p + (E/\zeta)] \qquad (7.33)$$

and then from equation 7.31

$$f^*(p) = [p + (E/\zeta)]/Ep. \qquad (7.34)$$

Hence one obtains

$$f(t) = \frac{1}{E} + \frac{t}{\zeta} \qquad (7.35)$$

Thus upon application of a constant stress $\sigma(0)$ the strain, as given by equations 7.26 and 7.35, is

$$\epsilon(t) = \sigma(0) \cdot \left(\frac{1}{E} + \frac{t}{\zeta} \right)$$

and is thus seen to consist of the contribution $\sigma(0)/E$ made by the Hookean element, and $\sigma(0).t/\zeta$ contributed by the Newtonian liquid, as was to be expected from elementary considerations.

With a parabolic creep law

$$f(t) = A.t^n, \qquad 0 < n < 1, \qquad A = \text{constant}$$

one finds similarly

$$r(t) = B \cdot t^{-n}, \quad B = (1/A) \cdot \frac{\sin n\pi}{n\pi}$$

8

LIQUIDS

8.1 Viscosity of Simple Liquids and Dilute Colloidal Suspensions

Newtonian behaviour in which the shear rate is proportional to the stress is found in liquids where the shear energy is dissipated mainly by collisions between small molecules. It is observed in water, solutions of inorganic salts, many oils, and in dilute colloidal solutions and suspensions. Suspended matter increases the viscosity. Provided its concentration c, expressed as volume fraction, does not exceed a few per cent the increase is represented by Einstein's formula

$$\eta = \eta_0 \, (1 + 2 \cdot 5c) \tag{8.1}$$

With high concentrations however η_0 is generally found to become dependent upon the shear rate.

Deviations from Newtonian behaviour may be grouped under two main headings. First we have liquids the rheological behaviour of which can be described fully in terms of the stress and the shear rate, and second, liquids which undergo structural changes in the course of shearing so that the viscosity becomes time-dependent even at constant shear rates. These effects may be reversible; the liquids return to their initial state as soon as the stress is removed or after a certain period at rest.

8.2 Non-Newtonian Liquids

Liquids displaying time-independent non-Newtonian behaviour can be grouped according to whether the shear rate increases with stress faster or slower than for a

122

Newtonian liquid. The relation between the stress and the shear rate then becomes non-linear. Many equations have been proposed empirically or semi-empirically to describe these phenomena. We shall here utilise only the empirical power-law representation, mainly because of its simplicity. One has

$$\tau = \eta_0 (d\gamma/dt)^{1+\delta} \qquad (8.2)$$

where η_0, the 'apparent viscosity' is a constant; it should be noted that its dimension here depend upon the index δ, which makes it rather difficult to ascribe a simple physical significance to η_0 in this representation of the flow.

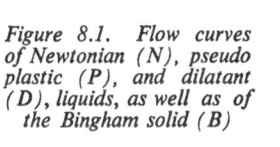
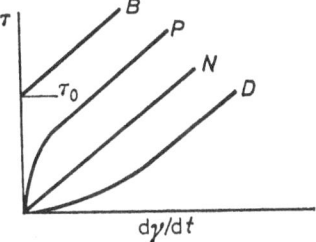

Figure 8.1. Flow curves of Newtonian (N), pseudo plastic (P), and dilatant (D), liquids, as well as of the Bingham solid (B)

If $-1 < \delta < 0$ the liquid is described as 'pseudoplastic' (*Figure 8.1*); if it were truly plastic the flow curve would have to show a definite yield point, e.g. as for the Bingham solid, also indicated in *Figure 8.1*. Liquids for which $\delta > 0$ are generally referred to as 'dilatant'.

Pseudoplastic behaviour is found in solutions of cellulose derivatives and other high polymers, and in suspensions containing elongated particles. These will tend to align themselves at high shear rates so that the long axes or dimensions lie along the stream lines, thus offering a reduced resistance to the shear flow. The most pronounced deviations from Newtonian behaviour will therefore occur at low strain rates.

Dilatancy is often found in concentrated suspensions,

such as fine sand in water, or lime-stone dust in bitumen. The packing of the particles becomes less close in rapid shear flow than while the material is at rest and settled, the suspension therefore swells as the shear rate is increased. Energy losses by particle collisions and attrition diminish at the same time and the liquid becomes more nearly Newtonian.

Paints, toothpaste and certain slurries often show Newtonian behaviour only if a certain yield stress τ_0 is exceeded. The rheological flow equation then becomes

$$\tau - \tau_0 = \mathrm{d}\gamma/\mathrm{d}t$$

Such materials are not, strictly speaking, liquids as they maintain their shape at stresses less than τ_0. The existence of the yield stress places them amongst ' plastic ' bodies; the simple model behaving as indicated by the line B in *Figure 8.1* was first proposed by Shvedov in 1889, and later studied in detail by Bingham, after whom it is named.

Time-dependent effects occur in thixotropic and rheopectic liquids. In the former a breakdown of the structure occurs on shearing, and the apparent viscosity decreases with increasing shear rate and length of time of shearing. The more linkages are broken the fewer remain intact, available for breakdown, so that at any given shear rate the viscosity attains a lower limiting value approximately exponentially with time. On standing, with the stress removed, the liquid eventually regains its original consistency. ' One-coat ' paints, lime pastes and clays, for example, exhibit this behaviour.

In rheopectic materials formation rather than degradation of structure occurs on continued slight agitation; large shearing movements inhibit the effect however. For example gypsum paste containing slightly over 40 per cent by volume of solid in the form of particles with a granulometry of about 1–10 μm, which would solidify in $\frac{1}{2}$–1 hour after shaking if left to stand, can be made to solidify in a few seconds if, after shaking, it is gently rocked in the container.

8.3 Elastic Liquids and the Weissenberg Effect

Viscoelasticity is generally exhibited by concentrated solutions of polymers and other macromolecular substances. A simple model is the Maxwell liquid discussed in Chapter 7. More complex behaviour can be represented by connecting at least one Voigt–Kelvin element in series with a dashpot or Maxwell liquid. The compound body would then display stress relaxation at constant strain and, on removal of the load it would contract somewhat, elastic

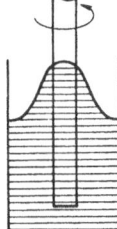

Figure 8.2. Climb of a viscoelastic liquid up a rotating rod, due to the Weissenberg effect

recovery taking place. The occurrence of either or both effects distinguishes a viscoelastic material from a purely viscous, not necessarily Newtonian, one. The elastic contribution to the deformation is responsible for a much discussed phenomenon, known as the Weissenberg effect. One of its manifestations is the climbing of liquid up an inner rod or cylinder of a coaxial system in relative rotation about the axis, as indicated in *Figure 8.2*. The effect does not depend upon the magnitude of the viscosity. It is readily demonstrated with, for example, a solution of a few per cent of polymethylmethacrylate in dimethylphtalate, polyisobutylene in dichlorobenzene, or aluminium laureate in paraffin.

125

The origin of the effect may be explained as follows. Consider the stress on a small element of thickness dz (*Figure 8.3*) of the rotating liquid, at a distance r from the axis of the system. We take a local set of axes, with σ_{xx}, σ_{yy} and σ_{zz} in the circumferential, radial and axial directions respectively. The stresses will then be as shown in the figure. If the liquid is at rest, i.e. not sheared, the above three stresses are all equal to $-p_0$, the hydrostatic stress at the point of location of the element under consideration.

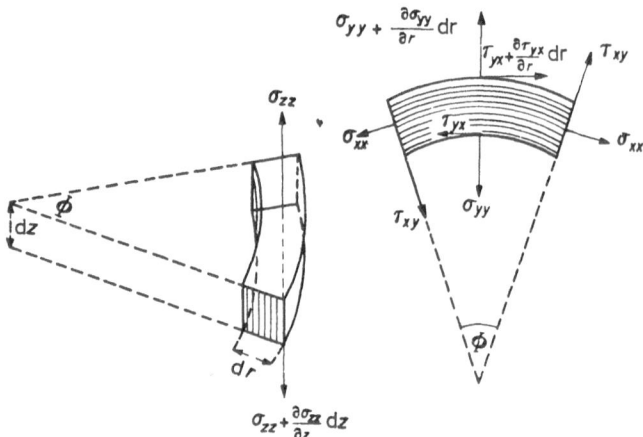

Figure 8.3. Stress distribution in a viscoelastic liquid in laminar shear flow

The sheared liquid may be regarded as bunched elastic fibres aligned, more or less, along the circumferential flow lines, immersed in a viscous, e.g. Newtonian liquid. The appropriate rheological model would then be a Maxwell body. The shear stress τ_{yx} inducing laminar flow will therefore also extend the 'spring element', in this case the fibres, subjecting them to a slight circumferential tension. Assuming the material to be incompressible, the

126

longitudinal extension of the fibres will be accompanied by radial and axial contractions, as if σ_{yy} and σ_{zz} had been changed by equal amounts, both to give a hydrostatic pressure greater than $-p_0$. The circumferential tension, and this effect, may be described formally by writing

$$\sigma_{xx} - \sigma_{yy} > 0 \qquad (8.3)$$

and

$$\sigma_{zz} - \sigma_{yy} = 0 \qquad (8.4)$$

Now, by considering the equilibrium of forces in the direction normal to the shearing planes one finds, from *Figure 8.3*

$$(\sigma_{yy} + \frac{\partial \sigma_{yy}}{\partial r}) (r + \delta r) \, \delta\phi \, \delta z - \sigma_{yy} r \, \delta\phi \, \delta z = \sigma_{xx} \, \delta r \, \delta z \, \sin \phi$$

which yields, as δr and $\delta\phi \rightarrow 0$

$$r \frac{\partial \sigma_{yy}}{\partial r} = \sigma_{xx} - \sigma_{yy} \qquad (8.5)$$

In view of equations 8.3 and 8.4 one can infer from equation 8.5 that

$$r \frac{\partial \sigma_{zz}}{\partial r} = \sigma_{xx} - \sigma_{yy} > 0 \qquad (8.6)$$

Since σ_{zz} is negative, i.e. a compressive stress, one can re-write equation 8.6 in the form

$$d \, | \, \sigma_{zz} \, | \, / dr < 0$$

This inequality implies that in the sheared liquid the absolute value of the hydrostatic pressure decreases as r increases; the pressure on the horizontal surface of the liquid therefore tends to be greatest near the inner cylinder or rod. In the course of rotation a dynamic equilibrium will consequently establish itself, with the higher pressure at the centre balanced by a higher column of liquid near the inner cylinder than at the walls of the container, as indicated in *Figure 8.2.* Equations 8.4 and 8.5 yield the relation

$$- \frac{d\sigma_{zz}}{d \ln r} = \sigma_{yy} - \sigma_{xx}$$

which, together with equation 8.4, has been used in experimental investigations of the Weissenberg effect.

BIBLIOGRAPHY

1. C. A. WERT and R. M. THOMSON, *Physics of Solids*, McGraw-Hill, New York, 1964

2. J. M. ZIMAN, *Principles of the Theory of Solids*, Cambridge University Press, Cambridge, 1964

3. S. TIMOSHENKO, *Theory of Elasticity*, McGraw-Hill, New York, 1934

4. L. R. G. TRELOAR, *The Physics of Rubber Elasticity*, Oxford University Press, London, 1958

5. C. ZENER, *Elasticity and Anelasticity*, University of Chicago Press, Chicago, 1948

6. J. C. JAEGER, *Elasticity, Fracture and Flow*, Methuen, London, 1956

7. A. NADAI, *Theory of the Flow and Fracture of Solids*, McGraw-Hill, New York, 1950

8. D. C. DRUCKER and J. J. GILMAN (Eds.), *Fracture of Solids*, Interscience, New York, 1963

9. F. R. EIRICH (Ed.), *Rheology I*, Academic Press, New York, 1956

10. B. PERSOZ (Ed.), *Introduction a l'Étude de la Rhéologie*, Dunod, Paris, 1960

11. R. HILL, *Plasticity*, Oxford University Press, London, 1950

12. W. OLSZAK, Z. MROZ and P. PERZYNA, *Recent Trends in the Development of the Theory of Plasticity*, Pergamon, Oxford, 1964

13. V. S. POSTNIKOV (Ed.), *Relaxation Effects in Metals and Alloys* (Relaksatsionnie Yavlenia v Metallach i Splavach), Metallurgizdat, Moscow, 1963

14. J. G. TWEEDDALE, *The Mechanical Properties of Metals*, Allen and Unwin, London, 1964

15. A. GELEJI, *Bildsame Formung der Metalle in Rechnung und Versuch*, Akademie Verlag, Berlin, 1960

16. V. A. PAVLOV, *Physical Principles of the Plastic Deformation of Metals* (Fizicheskye Osnovy Plasticheskoi Deformatsii Metallov), Academy of Sciences of the USSR, Moscow, 1962

17. J. J. GILMAN (Ed.), *The Art and Science of Growing Crystals*, Wiley, New York, 1963

18. H. B. HUNTINGTON, *The Elastic Constants of Crystals*, Academic Press, New York, 1958

129

DEFORMATION AND STRENGTH OF MATERIALS

19. C. R. EVANS, *An Introduction to Crystal Chemistry*, Cambridge University Press, Cambridge, 1964

20. P. H. GEIL, *Polymer Single Crystals*, Interscience, New York, 1963

21. J. FRIEDEL, *Dislocations*, Pergamon, Oxford, 1964

22. M. GORDON, *High Polymers*, Addison Wesley, Reading, Mass., 1964

23. C. KLINGSBERG (Ed.), *The Physics and Chemistry of Ceramics*, Gordon and Breach, New York, 1963

24. YA. FRENKEL, *Kinetic Theory of Liquids*, Oxford University Press, London, 1947

25. W. L. WILKINSON, *Non-Newtonian Fluids, Fluid Mixing and Heat Transfer*, Pergamon, London, 1960

26. A. S. LODGE, *Elastic Liquids*, Academic Press, New York, 1964

27. J. R. VAN WAZER, J. W. LYONS, K. Y. KIM and R. E. COLWELL, *Viscosity and Flow Measurements*, Interscience, New York, 1963

INDEX

Page numbers set in italic type denote most important entries.

INDEX

133